Praise for the Book

Emily Singler has provided a thorough resource for veterinary industry profession-als regarding the factors playing into the experience of pregnancy and early parent-ing. This is a highly valuable source of information that should be on the shelf of every single practice manager, OSHA officer, HR manager, medical director, prac-tice owner, department head, and more. It should be available on-hand for every single employee who may become pregnant, for easy timely reference.

On a more personal note, the information in this book is exactly what I was looking for when planning my first pregnancy as a new graduate from vet school. I foraged bits of information from countless resources, and was often left confused. I wish I had this book 10 years ago! It is the perfect resource for my doula clients now.

This is a must-have resource for every veterinary professional considering preg-nancy or becoming a parent. The information here offers clarity and compassion to women and families at a time of much uncertainty. I will be referencing it fre-quently as a childbirth educator who exclusively works with veterinarians and advocates for all parents within the veterinary industry.

Emily Taylor Yunker, DVM, VMRT, CVH, doula (The veterinarian doula)

This book is amazing – I wish I had something like this when I was having my kids! Dr. Emily Singer has written a comprehensive guide for any veterinary professional thinking about having a family. For every part of navigating pregnancy and the postpartum period, her book offers facts sprinkled with real life experiences from mothers with a range of experiences and from all parts of the veterinary profession. My personal experience was trying to piece these things together and figuring it out as I went and mostly on my own. Dr Singer has distilled so much useful infor-mation, advice, references, and recommendations - this is an essential resource for every veterinary practice and people navigating parenting in the veterinary profession.

Annie Wayne DVM, MPH, DACVECC and mother of 3

Dr. Singler has written an authoritative, first-of-its-kind book on pregnancy in vet-erinary medicine. The information is well-researched, and each chapter is comple-mented by true stories from veterinarians and veterinary technicians. This book should be read by anyone in the veterinary field who plans on getting pregnant or is pregnant, and by anyone who works with or manages pregnant team members. Most definitely a tremendous and much-needed resource in the veterinary field.

Lori Teller, DVM, DABVP (canine/feline), CVJ

Dr. Singler's book is a much-needed, practical and relatable reference on occupa-tional health and safety issues surrounding pregnancy and the postpartum period for veterinary professionals. With the trends in the profession toward a predomi-nantly female workforce, this guide provides useful resources not only to those con-cerned about how their own maternal health risks, but it is also helpful for those implementing workplace policies. While there are still many knowledge gaps in this field, this guide is a comprehensive and easy-to-follow reference for the veterinary profession.

Erin Luley, VMD, MPH, DACVP

When I began creating the field of parental leave consulting and coaching support in 2006, I dreamed that one day there would be specialty experts to provide targeted guidance to those in need. With this groundbreaking book Dr. Singler has made this dream a reality, and in doing so has given us a way to transform the veterinary field's approach to parenthood and careers.

Whether you're a veterinary professional contemplating or planning for parenthood, an employer or practice owner, a grad student, intern, resident, or simply someone managing individuals becoming parents in the veterinary profession, this book is your essential companion.

In this comprehensive guide, Dr. Singler walks you through every aspect of this intricate timeframe. From practical insights on casemaking to ensure leave benefits the entire team, to concise summaries of relevant laws based on your location, this invaluable resource empowers employees and employers (of all sizes) to implement and strategize around parental leave.

Beyond the professional realm, this book delves into deeply personal considerations and offers a roadmap to navigate the often challenging terrain of infertility and loss with expert guidance, and gives insight on recognizing and resourcing perinatal mood and anxiety disorders.

Buy this book today and gift yourself, and a generation of veterinarians, with the practical and emotional tools needed to embark on the journey to parenthood while nurturing a flourishing veterinary career. Pregnancy and Postpartum Considerations for the Veterinary Team is the transformative handbook you've been waiting for.

Amy Beacom, Ed.D

Founder and CEO, Center for Parental Leave Leadership
Author: The Parental Leave Playbook

Pregnancy and Postpartum Considerations for the Veterinary Team

Precautions often apply to pregnancy in any workplace, but being a vet in practice presents additional specific risks. There are concerns and uncertainty about potential hazards, from radiation and inhalant anesthesia exposure, to zoonoses, and the additional mental stress in a profession that already carries high suicide risk.

This book reviews considerations for professionals in clinical veterinary medicine (large and small animal) while pregnant and after giving birth. Veterinarian and veterinary writer, consultant, and mentor Dr. Emily Singler speaks directly to veterinary team members (veterinarians, technicians, CSRs, assistants, students) who are pregnant or plan to become pregnant. She delivers scientific information on the specific risks to the mother and baby that may be encountered during pregnancy while working in veterinary medicine, with some of her own and others' experiences to add perspective and humor.

The book also covers topics related to mental health challenges, announcing a pregnancy and planning for parental leave, navigating the fourth trimester, and returning to work. We hope that having read this book, veterinary professionals—whether pregnant or working with pregnant colleagues—will feel better supported and empowered to make informed decisions.

Pregnancy and Postpartum Considerations for the Veterinary Team

Emily Singler, VMD

CRC Press
Taylor & Francis Group
Boca Raton London New York

CRC Press is an imprint of the
Taylor & Francis Group, an **informa** business

First edition published 2024
By CRC Press
2385 NW Executive Center Drive, Suite 320, Boca Raton FL 33431

and by CRC Press
4 Park Square, Milton Park, Abingdon, Oxon, OX14 4RN

CRC Press is an imprint of Taylor & Francis Group, LLC
© 2024 Emily Singler, VMD

ISBN: 9781032524993 (hbk)
ISBN: 9781032524979 (pbk)
ISBN: 9781003406907 (ebk)

DOI: 10.1201/9781003406907

Typeset in Minion
by Deanta Global Publishing Services, Chennai, India

To Isa, William, Jacob, Elijah, and Levi: my proudest accomplishment and my greatest blessing is being your mama. Thanks for also giving me lots of interesting stories to include in this book!

To Daniel: thank you for being my forever companion on this journey … for trying to understand my craziness, for supporting me in my dreams, for helping me see the lighter side of life, and for being by my side in the trenches.

I love you all with all my heart.

Contents

Foreword

I first met Dr. Emily Singler virtually, during the pandemic, and was immediately impressed with her mission: to empower pregnant veterinary staff and their employers with the information they need to feel informed, supported, and safe at their workplace both during and after their pregnancy.

Dr. Singler and I quickly realized we have a lot in common, as we are part of a growing community of veterinarians who are also writers. Dr. Singler and I each founded and manage Facebook pages to support other veterinary writers: Community of Veterinary Writers (Dr. Singler) and Veterinary Narrative Medicine (myself and Dr. Annie Wayne). We also share a textbook publisher; my book *Narrative Medicine in Veterinary Practice: Improving Client Communication, Patient Care, and Veterinary Well-being* was published by CRC Press in 2021. I have also written a New York Times bestselling memoir, *The Other Family Doctor, a Veterinarian Explores What Animals Can Teach Us about Love, Life, and Mortality.*

As the mother of five children, Emily brings a wealth of lived experience to the subject of pregnancy and parenthood in the veterinary workplace. She has also included many personal stories from other parents to provide a broad spectrum of information and perspectives. Dr. Singler knows that "knowledge is power." The information and wisdom contained in these pages will help pregnant people make educated decisions for themselves and their families. Employers and practice managers will benefit greatly from this book as well, as it provides an invaluable resource for how to legally, safely, and practically accommodate and support pregnant employees and those with young children. The scope of the book is wide and includes information about fertility treatments, adoption, and post-partum issues such as childcare options and pumping at the workplace.

In addition to concerns a pregnant person may have about how their life will change after adding a young human to their family, in veterinary medicine specifically there may be numerous questions about the workplace environment. What are the risks to a pregnant person from anesthetics, radiation, chemicals, and infectious diseases, and how can they be mitigated? How can a pregnant employee discuss these matters with her employer and co-workers? How can an employer utilize best practices to keep their employees safe? How can a hospital team best support a pregnant co-worker, one who has suffered a miscarriage, or one who has terminated a pregnancy? Many obstetricians are unaware of the unique complexities involving veterinary practice employment. No longer will pregnant veterinarians and staff need to start from scratch in considering how they can work safely or wonder how other pregnant employees and practices deal with conditions such as drug and radiation exposure, ergonomic issues, and planning parental leave.

In my book *The Other Family Doctor*, I describe a disturbing conversation I had as a new graduate in the early 1990s. I was told by a prospective employer that he refused to hire a woman veterinarian, as he didn't want to bother training an associate "who would just get pregnant and quit." Although such views are less likely to be spoken aloud today, they may still be present and are not unique to veterinary medicine. This book will help to reduce discrimination by providing clear guidance to employers and practice managers. Now more than ever, this book is needed; with more women employed in veterinary clinics, pregnancy and child-rearing is a component of many people's lives. And as anyone with children knows, pregnant people and those with young children can and do function as valuable and dedicated personnel. In addition, parenting can help improve communication skills, patience, and the ability to create innovative solutions, as well as numerous other capabilities.

Pregnancy and Postpartum Considerations for the Veterinary Team will empower countless pregnant people and their families with the support and confidence to make personal decisions regarding pregnancy and returning to work as the parent of a young child. Dr. Singler has written the book I wished I had when I was pregnant: a thoroughly researched, comprehensive guide that belongs on the shelf of every veterinary clinic and in the hands of anyone contemplating the addition of a child to their family while working in the veterinary industry.

Karen Fine, DVM

Acknowledgments

This book would not be what it is without the help, guidance, and encouragement of family, friends, colleagues, other healthcare professionals, and fellow vet med moms. What started as a vague idea that was too "crazy" to pursue has turned into a resource that will hopefully help answer questions and guide new (and experienced) parents in veterinary medicine to make informed decisions during pregnancy and beyond.

I am very grateful to the first people who heard my idea and responded with the enthusiasm I needed to feel empowered and validated to bring this idea to life. These first people include my husband, Daniel Singler, my sisters Alice Remmey and Deborah DiNardo, and my friends and colleagues Maria Donnelly, DVM, MSPH, Melanie Barham, DVM, PMP, MBA, and Amanda Modes, DVM. Thank you for sharing my excitement and my vision.

As I started to work more intently, I felt the need to seek guidance from experts in some of the topics in the book. I am grateful to Carson Rodeffer, MD (Ob/Gyn), Nicole Wheeler, MD (Infectious Disease), Marie Holowaychuk, DVM (Veterinary Wellbeing), and Erin Luley, VMD, MPH, DACVP (Public Health and Occupational Health) for taking time out of their busy schedules to talk with me, answer my questions, and give me valuable insight during the research phase of my book. I sent a questionnaire out that was completed by several veterinarian and technician moms. I am grateful to each and every one of them, whether known or anonymous, who shared their triumphs and challenges with me and helped me make sure I was covering everything I needed to.

During my writing of the book, I was fortunate enough to find a wonderful editor, Andrea Winn, who showed incredible patience as I realized how much longer it would take for me to do this book justice while working full time and raising a family. She never complained as I asked her to push the timeline back eight times during the course of our work together. She guided me through so many parts of writing a book that were completely foreign to me, and she helped me make the book much better than it would have been otherwise.

I am endlessly thankful for having connected with Sonja Olson, DVM, and Karen Fine, DVM, fellow authors, my unofficial mentors, and wonderful, intelligent women. They helped me see what would be possible and cheered me on through my anxiety and imposter syndrome. I am so lucky to have been able to work with Alice Oven at CRC Press, who has been so kind, understanding, and helpful from the very moment we met.

Woven throughout the pages of this book are stories of mine, but also stories of many other veterinary professionals. They were gracious enough to share their happiest moments, their most painful experiences, and everything in between. I thank each and every one of them for their generosity and their time. I thank my friends and colleagues who reviewed my chapters, provided feedback, and pointed out what I couldn't see for myself after having read it so many times. I was never so happy to receive criticism, and I thank you for giving it.

Finally, thank you to my husband and children for giving up time with me while I sat on my computer during every free moment to write, edit, or work on other parts of the book. I've missed you, but I appreciate your support more than you know. Thank you to Daniel for taking a glimpse into my world and turning it into art. Thank you for going outside your artistic comfort zone and trying something new. Thank you for letting me be your "client" and for tolerating all of my requests to change things. Thank you learning a little about my experiences as a veterinarian, a woman, and a mother to help bring authenticity to your artwork.

About the Author

Dr. Emily Singler is a 2001 graduate of Penn State University and a 2005 graduate of University of Pennsylvania School of Veterinary Medicine. She is married with five kids and lives in Orlando, Florida. Her career in veterinary medicine has included experience in shelter medicine, private practice, and as a relief veterinarian. She currently works as a veterinary writer, consultant, and mentor and enjoys writing for both pet owners and veterinary professionals. Her writing interests include public health, preventive medicine, the human–animal bond, and life as a working mom. In her free time, she enjoys spending time with family, horseback riding, and all things llama and alpaca. More information can be found about her work at her website, www.emilysinglervmd.com and her blog, www.vetmedbaby.com.

About the Illustrator

Daniel Singler is an avid lover of art and nature, a rugged outdoorsman, a father of five, and just happens to be married to a veterinarian. He works as a vestibular physical therapist by day and enjoys spending time with his family and pursuing his other interests in his "spare time." On any given day, you might find him hiking and camping, painting a mural, trying out a new recipe in the kitchen, or crafting his own shoes out of leather. His artistic preferences have typically centered around landscapes and nature scenes, but he enjoys learning new techniques and experimenting with different media, including watercolors, graphite, and photography.

Introduction

Congratulations! I'm so glad you're here. Whether you're reading this book because you are already pregnant or planning a pregnancy, or whether you employ or work with others who are navigating their pregnancy journey, you are here to get information and to make smart and safe decisions. I hope to be able to provide you with the tools you need to keep yourself or those you work with safe and healthy while working in veterinary medicine.

When I found out I was pregnant with my first child, I had only been out of vet school for 3 months. I had just started my first job as a full-time associate veterinarian. I was married and very much planned on having children, but when I had that positive pregnancy test in front of me, I panicked. I didn't know what I was doing, how to be pregnant, or how to take care of a child. I was still trying to convince my clients that I was old enough to be a doctor! And I was worried that I would be fired because of my pregnancy. Looking back on that time now, I know that labor laws prevent firing based on pregnancy. But in my research for this book and in conversations with multiple experienced and successful veterinarians who are also parents, I know I was not the only one to experience this fear.

Thankfully in my five pregnancies, I never (knowingly) experienced discrimination at work. I continued working full time through all of them and was met with congratulations and understanding by my managers. I was able to take (largely unpaid) maternity leave. My worst fears proved to be unfounded. But I was also largely left to my own devices to determine what I should or should not be doing differently at work to protect myself and my unborn child. In many cases neither my employer nor my health care provider could provide me with much in terms of concrete guidelines or peer-reviewed evidence to inform my actions. (Most human health care professionals have limited familiarity with many of the hazards discussed in this book, including most zoonotic diseases, inhalant anesthesia exposure, having to actually restrain patients for radiation, etc.) So, I made things up as I went along based on my knowledge, some research, and what seemed like good common sense, often without anyone to consult with who had done it before me. I looked things up, I asked questions, and I informed my coworkers of why I was doing what I was doing. Did I sometimes make mistakes or outright bad choices? Of course! I shouldn't have tried to draw blood from a cat by myself when I was 37 weeks pregnant (or ever!). I ended up getting clawed in the face and bitten multiple times on my hand. I called my OB right away and she asked *me* which antibiotic to prescribe. Everything ended up being fine, but I still had the scabs on my face when I delivered my daughter.

I worked in small animal practice throughout each of my five pregnancies, and each time, I searched for information to guide my decisions so that I could best keep myself and my baby safe. Each time I came back from my searches disappointed with what I found. I wanted a book to help guide and support me, a sort of "what to expect when expecting and working in vet med practice," and it didn't exist. So that's what I'm hoping this book will be for you.

Before we delve any deeper, let's review what this book is not. First and foremost, this book is *not* to be construed as medical advice. I am not a human healthcare provider and cannot give advice about the management of a pregnancy or the treatment of any specific health conditions. All pregnant individuals and those trying to conceive should seek the regular care of a qualified medical professional and address any concerns about their health to them. The medical information presented here is meant to provide practical information about the risks that may exist in the practice of veterinary medicine so that you can make informed decisions and consult an appropriate medical professional for additional guidance. This can also not be construed as an all-inclusive list of every potential risk of working in veterinary medicine during pregnancy, although every effort has been made to be as comprehensive as possible. This book is also not a pregnancy book in the sense that it does not review the growth and development of an embryo or fetus or the changes in the pregnant mother. If you are so inclined, I would recommend purchasing such a book to read along during your pregnancy. This book, however, is specific to occupational hazards and other considerations relevant to working through pregnancy, navigating the postpartum period, and returning to work in veterinary medicine.

This book is also not meant to scare or discourage any veterinary professional from working during the conception, pregnancy, and early child rearing phases of life. We do discuss some scary possible sequelae of exposure to some workplace hazards, but always with the goal of informing you so that you can take preventive measures to reduce your risk. During my research for this book, I sought feedback from several veterinarians who have worked through their pregnancies and beyond, and when asked for advice for others planning to do the same, they all echoed the same sentiment—it can be done! Not only can it be done, but it can be very rewarding.

More than anything, this book is about taking care of you. You deserve to be protected from injury, pain, discomfort, sickness, stress, and more as you navigate your pregnancy and postpartum period both while working in veterinary medicine and while home healing and bonding with your baby. And if your child comes to you via gestational carrier or adoption, or if you are a non-birthing partner, this book is for you as well. If you are an employer, this book contains advice on how to best support all these individuals so that they can feel fulfilled as parents and successful in the workplace.

You will notice throughout this book that I use gender-neutral language where possible. This is in an effort to be as inclusive as possible. I recognize that not all people capable of becoming pregnant identify as women or female, and that not all non-birthing parents are men. I also recognize that some parents will be single by choice. I still use the terms mother and father in some instances, and where the references I quote use gender-specific language, I do not change it. I hope that all people who choose to read this book will be able to see themselves in the pages and find the support that they need. I have also included an appendix where you can find even more resources to support your journey.

Thank you for reading, and congratulations again on your journey!

Emily Singler, VMD

"Doc wants a straighter view!"

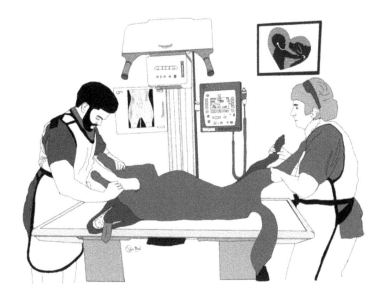

Radiation and imaging

When I was pregnant with my first baby, I was helping to position an animal on the X-ray table (not wearing any personal protective equipment because I planned to move before the exposure was taken). My techs didn't wait as I had instructed, though, and took the shot while I was standing right next to the table. I hadn't told anyone I was pregnant, and I was nervous about my accidental exposure. So, I panicked and blurted out that I was pregnant. As a first-time mom and brand-new vet, I knew that I should avoid radiation but didn't really know the specifics. I later learned that a single accidental exposure to scatter radiation carries a very low risk of harm to an embryo or fetus. Nevertheless, I committed to taking the necessary steps to limit my exposure to radiation in general and to encouraging my coworkers to do the same.

Imaging techniques including ionizing radiation, ultrasound (US), nuclear medicine imaging, and magnetic resonance imaging (MRI) are used in veterinary medicine to diagnose many different conditions in multiple species. Ionizing radiation can be divided into diagnostic radiography, fluoroscopy, computed tomography (CT), and radiation therapy for

DOI: 10.1201/9781003406907-1

1

the treatment of cancer. Nuclear medicine can be used both for diagnostic purposes and for treatment of hyperthyroidism in cats. Although US and MRI carry virtually no risk to a pregnant individual or fetus, varying degrees of risk have been identified with exposure to ionizing radiation and nuclear medicine. For the most part, typical occupational radiation doses are far below the doses that would be expected to cause complications during pregnancy. During lactation, only radioactive iodine (iobenguane I 131) presents potential risks of harm. Most risks can be greatly reduced or eliminated by understanding and appreciating their magnitude and following appropriate safety precautions.

IONIZING RADIATION

DIAGNOSTIC RADIOGRAPHY

X-ray and CT are among the most well-known hazards to a developing baby that veterinary and other healthcare professionals frequently encounter. Unlike in human medicine, in which most healthcare providers are usually completely shielded from radiation by a physical barrier, or not present at all for the exposure, veterinary technicians and veterinarians often restrain their patient for X-rays and are thus more likely to be exposed to scatter radiation. This is a concern for team members who are or may be pregnant because rapidly dividing cells are very susceptible to damage from radiation. With proper planning and precautions, the risks of exposure can be greatly reduced or eliminated altogether.

All humans are exposed to a low level of background radiation from the sun and stars, various elements in the earth, and food sources. According to the American College of Obstetricians and Gynecologists (ACOG), the relative effective dose of radiation to which humans are exposed, defined as the "amount of energy deposited per kilogram of tissue normalized for biological effectiveness," is commonly measured in milliGrays (mGy).[1] The estimated average exposure to radiation for a human being is about 2 mGy per year.[2] However, the radiation dose to which a fetus is exposed is generally lower than that of the pregnant individual because the uterus and other nearby structures help to shield the fetus.[3] A list of estimated fetal radiation doses, according to ACOG and the International Atomic Energy Agency (IAEA), after various types of radiation in which a pregnant person is undergoing diagnostic imaging, is shown in Table 1.1.[1,4,5]

While these are the estimated doses of radiation that could reach the embryo or fetus of a pregnant patient who is the *primary subject* of radiation exposure and who is positioned in the primary beam, occupational exposures to any pregnant personnel assisting with the imaging should be considerably less if proper safety precautions are observed. Furthermore, the radiation dose received by the embryo or fetus of a healthcare provider assisting with radiation exposures will be even lower because of the shielding properties of the uterus and other surrounding tissues.

RADIATION THERAPY

Radiation therapy for the treatment of cancers in humans and animals involves very high doses of radiation to kill rapidly dividing tumor cells. Total doses (to the patient, not the

Table 1.1 Estimated fetal radiation doses by type of radiation

Type of radiation	Estimated fetal dose (mGy)
Background radiation throughout a pregnancy	1
Dental radiographs	0.000009–0.00797
2-view thoracic radiographs	0.0005–0.01
Head or neck CT	0.001–0.01
2-view mammogram	0.001–0.01
Abdominal radiographs	0.1–3.0
Double-contrast barium enema	1.0–20
Abdominal CT	1.3–35
Fluoroscopic barium upper GI study	1.1–5.8
Fluoroscopic barium enema	6.8–24
Pelvic CT	10–50
Technetium-99m bone scintigraphy	4–5
PET whole-body scintigraphy	10–50

CT, computed tomography; GI, gastrointestinal.

fetal dose) can range from 45 to 60 Gy divided into smaller (3 Gy) daily doses 5 days a week for up to 2 months.[6,7] It is expected that fetal doses from radiation therapy would be considerably higher than those listed in Table 1.1, but this would depend on the part of the body being irradiated.

Noncancer effects

Both the CDC and ACOG report on the potential effects of acute exposure (a dose delivered at one time) of radiation to the embryo or fetus. Neither organization reports data on the differences in sequelae between acute (a single exposure) and chronic (multiple exposures over a period of time) exposures of the same total magnitude, except to acknowledge that there may be differences between the outcomes based on whether the exposure was acute or chronic.[3] Table 1.2 summarizes some of the potential effects of radiation exposure based on gestational age at the time of the exposure and the estimated fetal radiation dose.[1,3]

According to the CDC and ACOG, an embryo or fetus is sensitive to radiation damage with an exposure of at least 50–100 mGy.[1,3] Exposure greater than or equal to 500 mGy can result in severe effects, depending on the dose and the stage of pregnancy during which the exposure occurs.[3] It is worth noting that even the very lowest acute exposure thought to be capable of causing harm to an embryo or fetus is much higher than most of the estimated fetal radiation doses shown in Table 1.1.

The first two weeks after conception are a critical time for vulnerability of the embryo to radiation exposure. This is largely because the embryo consists of a very small number of cells at this point, and death or damage to any of those cells would carry a much higher risk of death of the embryo. It is suggested that doses of 50–100 mGy could result in death of the embryo during the first two weeks of pregnancy, and doses greater than 1000 mGy

Table 1.2 Potential effects of fetal radiation exposure based on gestational period and fetal radiation dose

Period of gestation	Fetal radiation dose	Potential effects
0–2 weeks	50–100 mGy	Possible death of embryo
0–2 weeks	>1000 mGy	High risk of death of embryo
2–8 weeks	>200 mGy	Congenital abnormalities, growth restriction
8–15 weeks	60–310 mGy	High risk of severe Intellectual disability
8–15 weeks	>200 mGy	Microcephaly
16–25 weeks	250–280 mGy	Low risk of severe intellectual disability
Any age	10–20 mGy	Possible increased risk of childhood leukemia
Any age	>100 mGy	Possible increased risk of cancer later in life

mGy, milliGray

would almost definitely be lethal to an embryo at this stage.[1] If an embryo survives, however, there are likely to be no negative consequences.[3]

From weeks two to eight after conception, doses greater than 200 mGy can result in congenital abnormalities and/or growth restriction. From weeks 8 to 15, it is theorized that doses ranging from 60 to 310 mGy could result in a high risk of severe intellectual disability, and that an exposure of 200 mGy could result in microcephaly (abnormally small head with a potentially underdeveloped brain). With much larger doses of radiation, there is an estimated 25 IQ point loss per 1000 mGy of exposure. From weeks 16 to 25, doses ranging from 250 to 280 mGy carry a low risk of severe intellectual disability.[1] After 24 weeks of pregnancy, no noncancer effects are expected from fetal radiation doses that do not exceed 500 mGy.[3] There is no risk of harm to a breastfeeding child whose mother is exposed to radiation while lactating.[1]

It is important to note that much of the information that is known or theorized about radiation exposure to human embryos and fetuses is derived from either animal studies, studies monitoring patients who required radiation therapy at higher doses for medical reasons, and studies of the survivors of the atomic bombs dropped on Japan in World War II.[1] For obvious reasons, no trials will ever be conducted specifically to measure the effects of radiation on developing human embryos and fetuses, so many guidelines are extrapolated from the available information. There are likely instances in which the theoretical risk does not completely correlate with the true risk in practice. However, given the potentially serious consequences associated with the risk, it is wise to mitigate the risk as much as possible.

CANCER EFFECTS

There is a small potential increased risk of cancer in children who were exposed to significant doses of radiation in utero. Fetal exposure of 10–20 mGy (which is still much higher than would be expected to occur with occupational exposure) may increase the risk of leukemia in children by 1.5 to 2 times the background incidence of this disease, which is approximately 1 in 3000.[1] Radiation doses greater than 100 mGy in utero may increase the risk of cancer for that individual later in life. Animal studies and the aforementioned study of atomic bomb survivors in Japan also point toward a theoretical tenfold increase in risk for adult-onset cancer before 50 years of age after radiation exposure during early childhood (from birth to 5 years of age) when compared with the risk after prenatal radiation exposure.[1] In other words, the risk for radiation-induced cancer later in life appears to be significantly lower for a radiation-exposed embryo or fetus compared with the risk for a radiation-exposed young child.

BOX 1.1   Practice profile: Avoiding radiation exposure

Amanda Modes, DVM

I got pregnant with my first son right around the time that I started a new job. Around six weeks (I'm not even sure if I had been to the doctor for confirmation yet) I was working an overnight emergency shift and we had a technician call out. This was a small ER so it was only me and two technicians in the building on an overnight shift. With one tech calling out, it was just me and one tech on staff that night, which meant I would have to step up and help with technician duties more than usual. We had a case come in that required radiographs. I had not told anyone at work that I was pregnant yet, but I was not comfortable taking radiographs either. I had to decide how I was going to handle this situation. I could tell the tech I was pregnant and unable to take radiographs. I could tell her I could not take radiographs and not explain why. Or I could avoid the conversation and take the risk and just take the radiographs. I felt I had to stand up for myself and would not take the radiographs, but I also felt the right thing to do was to explain the situation to the technician. Thankfully she was very understanding and agreed not to discuss it with anyone else until I announced to the rest of the office on my own time. As for the patient, the tech agreed to try to take the radiographs herself and if it was not going to work, we would decide what was best for the patient (sedation or referral).

RISK MITIGATION

AVOIDANCE

Pregnant veterinary personnel may elect to completely avoid restraining patients or otherwise being present during ionizing radiation exposures. This can involve stepping out of the room, around a corner, or behind a lead barrier. In some cases, complete avoidance of work with ionizing radiation is not practical. If veterinary personnel elect to continue to take radiographic exposures during pregnancy, they can follow several strategies to reduce the radiation dose to which they are exposed. These techniques are recommended for all personnel, regardless of pregnancy status, to reduce the risk of any adverse health outcomes.

DOSE REDUCTION

To reduce the radiation exposure, it is helpful to remember the acronym ALARA (as low as reasonably achievable). Strategies that help to accomplish this goal are centered around distance, time, and shielding.[8]

The risk of scatter radiation exposure can be decreased by increasing the distance between personnel and the source of the radiation. The recommendation is to stand at least six feet (about two meters) away from the primary beam, and farther away if possible. This is often accomplished by using sedation to avoid having to physically restrain the patient.[9] Positioning aids such as sandbags, foam pads, tape, and gauze can be used to manipulate and immobilize the body part in question with less need for physical restraint. For those who cannot maintain six feet of distance from the primary beam, care should be taken to avoid having any body part, shielded or not, in the primary beam.[8] Many dental radiography units also recommend that personnel maintain a six-foot distance from the beam and the patient during exposures to reduce the dose of radiation.[9]

Radiation dose can also be reduced by taking fewer exposures. The use of technique charts can aid in the proper positioning and exposure settings for common views based on the equipment specifications and patient anatomy and size. Collimating the beam to include only the body part in need of evaluation will reduce scatter and produce a higher quality image. Using the image processing software associated with digital radiographs to adjust contrast and further analyze images can also decrease the need to repeat exposures. All of these techniques can also result in less time spent in the radiology suite, which reduces the occupational radiation dose to personel.[10]

Anyone involved in the restraint of a patient should wear personal protective equipment (PPE), including lead aprons, gloves, thyroid shield, and goggles. These are meant to protect from scatter radiation and not the primary beam.[8] PPE is also necessary when taking dental radiographs unless the operator can stand at least six feet away or behind a permanent barrier.[10] Permanent barriers such as lead screens provide even better protection and reduce the need for PPE in many cases.[8]

BOX 1.2   Practice profile: Radiation safety during pregnancy

Amanda Shull, DVM

As a pregnant woman, my experience with radiology safety at work was less than ideal.

The building where I practice has the X-ray area directly next to the surgery room. The screen for viewing X-rays is located within the surgery room. The wall between the X-ray area and the surgery room does not contain any barrier (lead or otherwise).

Our schedule is such that each doctor performs surgeries usually one day per week. On my surgery day, I would have to stop in the middle of procedures to leave the room to allow for other doctors to have radiographs performed on their patients. Additionally, another doctor in the practice would instruct the assistants and technicians to "go ahead" with X-rays without giving me a heads up to leave the area. Several technicians went to bat for me and reminded my colleague of the risks to me and my baby. This colleague made no changes, and I was forced to go to management multiple times to correct the issue. Finally, the technicians and assistants refused to take radiographs for anyone when I was in surgery unless it was an emergency.

LARGE ANIMAL AND AMBULATORY RADIOGRAPHY

There are some unique differences between small animal radiography and large animal and/or ambulatory radiography that can require additional precautions to help reduce the risk of exposure for pregnant personnel. While some large animal radiographs are taken in a controlled, fixed environment, many are taken on the farm using a portable unit. It is also common practice for mobile practitioners to utilize the assistance of the animal owner or handler when a technician or assistant is not available. Another difference between large and small animal radiography is the direction in which the beam is oriented. While small animal radiography typically has the beam directed toward the ground, large animal radiography typically involves a beam that is oriented horizontally. Whereas small animal radiography typically has a fixed location for the cassette, large animal radiography usually requires that the cassette be held in place, increasing the risk of exposure for the individual holding the cassette.

The risks of exposure can be reduced using a combination of environmental control, proper positioning aids and restraint, and additional equipment that protects personnel. Portable X-ray devices should ideally be mounted on a stand, as opposed to being held, to allow the operator to stand at a distance when taking the exposure. The beam should never be directed toward people or toward areas where people could enter. Large animals should be properly restrained using ropes, stocks, and other means, with sedation as indicated. A sedated horse may benefit from having a head stand to rest their head on. Blocks can be used to raise a horse's foot off the ground for better centering of the beam on the foot without anyone having to hold the foot. All these measures can reduce the need for veterinary personnel to have to stand in close proximity to the patient and the beam. Using a plate holder on a long handle can also allow personnel or animal handlers to create more distance between themselves and the beam while holding the cassette in place. Appropriate PPE should be used by all individuals associated with radiography, whether it is in a radiology facility or outside on a farm.[8]

BOX 1.3 Practice profile: Equine radiography adaptations during pregnancy

Stacey Cordivano, DVM

AS TOLD TO EMILY SINGLER

Dr. Stacey Cordivano is a mom of two boys, equine veterinarian, practice owner, host of The Whole Veterinarian podcast, and leader in the movement to improve equine veterinarian retention and the overall happiness and wellness of veterinarians. She has owned her own practice for 12 years, which included both of her pregnancies. As an ambulatory lameness veterinarian, she averages about 3 to 5 X-ray cases per week, each of which can include anywhere from 4 to 16 images.

During both pregnancies, she elected not to take part in taking radiographs, although she knows of other equine veterinarians who continued to take radiographs during their pregnancies. She discussed the decision with her OB, who didn't really give her much feedback either way. She admits that she didn't really know what the actual risk was to her fetus, but she elected to just limit her exposure as a precaution.

She was fortunate to have a technician with her during her pregnancies who took on the role of taking the exposure, while the horse owner would hold the cassette. Stacey would go around the corner (she remembered that she should be at least six feet away). Both technician and owner would wear lead gowns, and whoever held the cassette would also wear lead gloves and try to stand off to the side. In some instances, she had to share a technician with another equine practitioner in the area.

For the few cases where she didn't have a technician with her, Stacey would double gown (one lead gown covering her front, one lead gown covering her back), and she felt comfortable doing this for the few times she had to do it. She has found that her X-ray badge hardly ever registers any exposure.

Stacey feels that the most important improvements that can be made from a safety perspective for pregnant ambulatory equine practitioners are increased usage of trained veterinary technicians on their calls, the ability to refer cases that pose a higher risk of injury to her colleagues (in her case, having to administer hindlimb injections), and having open communication with clients.

DOSIMETRY

Pregnant personnel who continue to take radiographic exposures must wear two dosimetry badges: a whole-body dosimetry badge that is worn outside of their PPE, and a second fetal dosimetry badge that is worn underneath the lead gown at the level of the abdomen. The whole-body dosimetry badge should not register more than 0.55 mGy in any given month, and special caution should be taken to avoid large doses of radiation between 8 weeks and 15 weeks of pregnancy, since the embryo is most sensitive to the effects of radiation during this time.[10] The fetal dosimetry badge must be checked monthly to ensure that the fetal radiation dose is within acceptable limits. The estimated fetal radiation dose (which will be lower than what is measured by either dosimeter) should be limited to 1 mGy throughout the entire pregnancy.[4]

FETAL RADIATION DOSE ESTIMATION

When there is a concern about the potential radiation dose that an embryo or fetus might receive or may have already received, health and medical physicists can help. Medical physicists study how radiation affects the human body and often work in hospitals and other human healthcare settings. They can help advise healthcare providers as to the safety of a radiographic procedure for a pregnant individual, and they can assess the severity of an exposure that has already occurred. Health physicists help to make sure radiation is used safely, for the benefit of both the public and the personnel who use it. While these experts generally do not work in veterinary hospitals, they will likely be present in university settings and governmental institutions, and they may be available for consultations with those who work outside of these settings. They can be a source of valuable information if there is a concern about radiation exposure during pregnancy.[11]

RADIATION THERAPY AND CT

Both radiation therapy and CT scans administer larger doses of radiation to a patient and therefore have the potential to pose a greater risk to healthcare workers as well. However, veterinary patients are almost always sedated or anesthetized for such procedures.[12,13] Because the sedation or anesthesia mostly eliminates the need for the physical restraint of small animals, there is rarely a reason for any veterinary personnel to be exposed, regardless of their pregnancy status. Large animals such as horses may still require a handler to be present, even when sedation is used. Handlers should wear PPE as described previously and stand as far away as possible from the patient and the radiation source to limit their radiation exposure, and pregnant personnel should ideally avoid any exposure to these modalities.[8]

NUCLEAR MEDICINE AND IMAGING

IOBENGUANE I 131

Iobenguane I 131 is a radioactive iodine isotope used for the treatment of hyperthyroidism in some cats. It is also used in humans as a diagnostic agent, given along with potassium iodide to block uptake into the patient's thyroid gland. In their recommendations to practitioners, ACOG states that I 131 should not be given to pregnant patients because it easily crosses the placenta and can negatively affect the embryo or fetus's thyroid gland. Pregnant individuals should avoid handling I 131, and any patients treated with it, in veterinary practice as well.[1]

I 131 is also excreted in breastmilk, where it can be absorbed and concentrated into a nursing infant's thyroid gland.[12] In human medicine, lactating mothers who need this diagnostic agent are instructed to not breastfeed for 80 days after the agent was last administered, which essentially means discontinuing breastfeeding altogether. Nursing mothers are advised not to work with substances containing I 131 in their workplace, which would also mean not working with I 131-treated patients.[14]

NUCLEAR MEDICINE IMAGING

Nuclear medicine imaging consists of nuclear scintigraphy and positron emission tomography (PET scan), both of which are infrequently used in veterinary medicine when compared to human medicine. A gamma ray-emitting radioisotope (such as technetium-99, indium, thallium, or radioactive iodine) is administered to a patient by injection, inhalation, or ingestion, and a gamma camera is used to detect the radioisotope in the body. Patients who undergo this type of testing typically stay at the imaging location for several days until they are less radioactive.[15]

The International Atomic Energy Agency (IAEA) reports that apart from iobenguane I 131, pregnant personnel can generally continue their work in nuclear medicine with some precautions.[16] These precautions include limiting time spent in the radiopharmacy, wearing appropriate PPE (which varies by procedure) at all times, preventing any skin contact with radionuclides, and limiting close contact with patients. All personnel should wash their hands after working in nuclear medicine. No food, drink, cosmetics, smoking materials, or any other personal items should be brought into the radiopharmacy, stored in a refrigerator with unsealed radioisotopes, or used while working in nuclear medicine.[8]

ULTRASOUND (US) AND MAGNETIC RESONANCE IMAGING (MRI)

According to ACOG, there is no recognized risk to the embryo or fetus with either US or MRI when it is medically necessary in pregnant individuals.[1] Therefore, no additional workplace precautions are recommended for pregnant veterinary personnel.

TAKE-HOME POINTS

- The acute fetal radiation doses associated with typical diagnostic imaging studies in veterinary medicine are typically well below the minimum doses for an adverse pregnancy event to be likely.
- Ionizing radiation represents no risk to lactating personnel or their infants.

- Pregnant team members may choose to avoid radiation exposure altogether during their pregnancy.
- When ionizing radiation exposure cannot be completely avoided, the ALARA principles can reduce exposure through increased distance, decreased time, and appropriate shielding.
- Large animal and ambulatory radiography can be made safer by using stands, plate holders, sedation, physical restraints, and positioning aids.
- Pregnant personnel who work with radiation should monitor their fetal radiation dose monthly using a second fetal radiation badge.
- MRI and US represent no risk to pregnant employees.
- The risk with CT scans and radiation therapy is very low because of the routine use of sedation or anesthesia for veterinary patients.
- Iobenguane I 131 and patients treated with it should be completely avoided by pregnant and lactating personnel.

REFERENCES

1. American College of Obstetricians and Gynecologists. Guidelines for diagnostic imaging during pregnancy and lactation. October 2017. Accessed February 18, 2022. https://www.acog.org/clinical/clinical-guidance/committee-opinion/articles/2017/10/guidelines-for-diagnostic-imaging-during-pregnancy-and-lactation
2. McLain Madsen L. Pregnancy in the workplace. February 18, 2022. https://www.vetfolio.com/learn/article/pregnancy-in-the-workplace
3. Centers for Disease Control and Prevention. Radiation and pregnancy: A fact sheet for clinicians. June 5, 2020. Accessed February 18, 2022. https://www.cdc.gov/nceh/radiation/emergencies/prenatalphysician.htm
4. International Atomic Energy Agency. Radiation protection of pregnant women in radiology. Accessed March 30, 2022. https://www.iaea.org/resources/rpop/health-professionals/radiology/pregnant-women#9
5. International Atomic Energy Agency. Radiation protection of pregnant women in dental radiology. Accessed February 17, 2023. https://www.iaea.org/resources/rpop/health-professionals/dentistry/pregnant-women
6. Thrall DE. Present status of radiation therapy in veterinary medicine. Paper presented at: World Small Animal Veterinary Association World Congress Proceedings; 2004; Raleigh, NC. Accessed February 18, 2022. https://www.vin.com/apputil/content/defaultadv1.aspx?id=3852189&pid=11181&
7. Smith Y. Radiation therapy dosage. March 2021. Accessed March 29, 2022. https://www.news-medical.net/health/Radiation-Therapy-Dosage.aspx
8. International Atomic Energy Agency. Radiation protection and safety in veterinary medicine. 2021. Safety Reports Series Number 104. Accessed March 29, 2022. https://www-pub.iaea.org/MTCD/Publications/PDF/PUB1894_web.pdf
9. Midmark. VetPro DC animal health dental x-ray system user manual. October 2019. Accessed March 28, 2022. https://technicallibrary.midmark.com/PDF/00-02-1601%20rev%20L01%20VetPro%20DC%20User%20Manual%20-%20English.pdf

10. Idexx Veterinary Radiation Safety Center. Why does radiation safety matter? Accessed March 28, 2022. https://www.idexx.com/en/veterinary/diagnostic-imaging-telemedicine-consultants/radiation-safety-center/

11. Occupational Outlook Quarterly. 2011. Medical physicists and health physicists: Radiation occupations. Accessed February 17, 2023. https://www.bls.gov/career-outlook/2011/summer/art02.pdf

12. University of Florida College of Veterinary Medicine Small Animal Hospital. Radiation treatment. Accessed March 28, 2022. https://smallanimal.vethospital.ufl.edu/clinical-services/oncology/oncology-faqs/radiation-treatment/

13. University of Florida College of Veterinary Medicine Large Animal Hospital. Computed tomography. Accessed March 28, 2022. https://largeanimal.vethospitals.ufl.edu/hospital-services/diagnostic-imaging/computed-tomography-cat-scan-or-ct/

14. National Center for Biotechnology Information. Drugs and lactation database (LactMed). Iobenguane I 131. August 17, 2020. Accessed February 22, 2022. https://www.ncbi.nlm.nih.gov/books/NBK501710/

15. Lattimer J. Nuclear medicine imaging in animals. November 2022. Accessed February 17, 2023. https://www.merckvetmanual.com/clinical-pathology-and-procedures/diagnostic-imaging/nuclear-medicine-imaging-in-animals

16. International Atomic Energy Agency. Radiation protection of pregnant women in nuclear medicine. Accessed February 17, 2023. https://www.iaea.org/resources/rpop/health-professionals/nuclear-medicine/pregnant-women

2

"I don't *smell* any iso!"

Anesthesia

When I was pregnant with my first child, I saw a nurse practitioner for my early appointments. She had no idea what isoflurane was, and when she looked it up, the warning described it as fetotoxic. She on the spot advised me to limit my anesthesia exposure to once a week, which seemed arbitrary. During a subsequent pregnancy, my OB also didn't seem to know how to advise me when I asked about continuing to perform surgery during my pregnancy. She ultimately remarked that it should be fine. I continued to perform surgery during all my pregnancies, with some precautions. Other pregnant veterinary personnel decide to avoid inhalant anesthesia exposure altogether or to wear a respirator. While there are potential risks associated with inhalant anesthesia exposure, those risks can be mitigated greatly, and the decision to continue with exposure to anesthesia during pregnancy is a very personal one.

There tends to be a fair amount of confusion about the safety of inhalant anesthesia exposure during pregnancy. A review of the literature reveals conflicting conclusions about the risk for spontaneous abortion, fetal death, congenital abnormalities, and/or low birth

DOI: 10.1201/9781003406907-2

weight. This can explain why many pregnant veterinary personnel are unsure of how to proceed at work while still minimizing the risk to their pregnancies. The good news is that most reports of adverse outcomes in pregnant healthcare workers are linked to high-exposure concentrations, either from improper scavenging or excessive leakage of waste anesthetic gas. Although there are some real risks associated with occupation anesthesia exposure, advancements in workplace safety over the past 40 years have helped to reduce the risk substantially. Anesthetic scavenging, for example, along with other workplace controls, greatly reduces the level of exposure to all veterinary personnel. Many veterinary personnel continue to perform surgery and utilize inhalant anesthetics for their patients throughout their pregnancies without incident. For those who feel comfortable continue working around anesthesia, it is important to understand the risks and the proper safeguards needed to protect all members of the team.

EVIDENCE OF RISK TO PREGNANCY FROM OCCUPATIONAL ANESTHESIA EXPOSURE

Studies that have attempted to determine if an association exists between occupational inhalant anesthesia exposure and adverse fertility and/or pregnancy outcomes have yielded mixed results. One of the earliest studies to look at the risk of anesthesia exposure in pregnant women was a small Russian study in 1967.[1] In this study, 31 pregnant anesthesiologists were evaluated. The women in the study had been exposed to inhalant anesthesia in an unscavenged operating room. Of those 31 pregnancies, there were 18 spontaneous abortions, 2 premature deliveries, and 1 congenital abnormality.[1] More recently, a 2008 survey of 940 pregnancies reported by Australian veterinarians found a statistically significant increased risk of spontaneous abortions in pregnant veterinarians who self-reported having 1 hour or more per week of exposure to unscavenged waste anesthetic gases.[2]

Chronic, low-level waste anesthetic gas exposure during pregnancy has been suggested to cause an increased risk of fetal genetic mutations including heart disease, cancer, kidney and liver disease, nervous system disorders, and reproductive disorders.[3] One study from 1974 suggests that the occupational anesthesia exposure of male personnel in an unscavenged operating room was associated with a higher risk of congenital abnormalities in their children even when their pregnant female partners had no anesthesia exposure.[4] Animal studies have also shown nerve cell damage, impairment of learning and memory, reduced sperm health and production, and birth defects after fetal isoflurane exposure.[5]

Although these studies suggest a concerning association between occupational inhalant anesthesia exposure and adverse pregnancy outcomes, there were likely flaws in study design in many cases, due to a lack of a control group to account for other variables such as stress, contagious disease exposure, or long hours. There were likely inconsistencies introduced by having participants self-report without having objective measurements of waste anesthetic gas levels. The lack of a proper scavenging system in these and other similar studies also likely resulted in much larger exposures than would be expected in current day workplaces.

BENEFITS OF SCAVENGING SYSTEMS

One of the biggest advances in the safety of anesthetic gas use for healthcare personnel is the anesthesia scavenging system. Scavenging can be done actively, in which the gases are removed by suction and then deposited outside the building in an area where they are unlikely to be taken back inside easily; or passively, where suction is not used and gases are adsorbed into a canister filled with activated charcoal. The Occupational Health and Safety Administration does not have specific requirements for the scavenging of anesthetic gases in the workplace, but it does have a "general duty clause" that mandates that employers provide a safe work environment for employees.[3] A scavenging system that is properly maintained and functioning can reduce the concentrations of waste anesthetic gases in the air by up to 90%.[3]

Proof of the benefits of a scavenging system can be found in studies that have been conducted since scavenging systems have become commonplace. A 1987 study measuring the anesthetic gas exposure of female veterinarians and technicians and correlating it with reproductive outcomes found no statistically significant increase in adverse events when pregnant personnel were exposed at levels near National Institute for Occupational Safety and Health (NIOSH)-recommended standards.[6] A 1990 study compared female graduates of a US veterinary school with female graduates from a US law school and found no statistically significant difference between rates of spontaneous abortion or low-birth-weight infants.[7]

Another study prospectively compared pregnant women with veterinary occupational anesthesia exposure to age- and gestational age–matched pregnant women with no occupational anesthesia exposure. No major differences were found between rates of spontaneous abortion or birth defects between the two populations.[8] A systematic review of studies looking at the health effects of occupational anesthesia exposure concluded that there was "no evidence of adverse effects when environmental levels are kept within legal threshold values."[9] The American College of Obstetricians and Gynecologists (ACOG), reaffirms, "No currently used anesthetic agents have been shown to have any teratogenic (birth defect causing) effects in humans when using standard concentrations at any gestational age."[10]

Although this cannot be interpreted as the perceived absence of risk for the developing baby, ACOG also recommends that medically necessary surgery and general anesthesia not be withheld from pregnant people at any gestational stage.[10] The limited studies of in utero exposure of a human fetus to anesthesia or sedation administered during pregnancy have not revealed any increased risk of brain development abnormalities that were linked to their exposure.[11]

RECOMMENDED EXPOSURE LIMITS

NIOSH has recommended a maximum exposure to inhalant anesthetics for all healthcare workers (pregnant or not) expressed as the concentration of halogenated anesthetic (which includes isoflurane, sevoflurane, and desflurane) in room air to be 2 parts per million (ppm).[5] If no precautions are taken, a veterinary team member's exposure may easily exceed this limit. The air around the nose of a patient with an inflated endotracheal tube cuff typically has an inhalant anesthetic concentration of 3 ppm. If the cuff is not properly inflated, the air concentration around the patient's face rises to 6 ppm. The concentration near an induction chamber is approximately 10 ppm. A recovering patient who is no longer intubated can have a concentration of 5 ppm near their face.[12] In an investigation of isoflurane concentrations in the veterinary technicians' breathing zone during an anesthetic procedure for a dog, baseline levels were just under 2 ppm. However, the reading spiked to almost 70 ppm when the endotracheal tube was disconnected from the breathing circuit while the dog was turned over.[5] These findings reveal the importance of exercising appropriate precautions when working around inhalant anesthesia and anesthetized patients.

PREVENTION

In most cases and with a properly functioning scavenger system, the risk of any kind of negative consequence to a developing embryo or fetus from anesthetic exposure during pregnancy is likely to be very low. However, there is still a need to take appropriate precautions to ensure that this is the case. Below is a discussion of ways to limit or avoid exposure to high levels of waste anesthetic gases.

AVOIDANCE

Some veterinary personnel will elect to avoid any anesthetic gas exposure during their pregnancy. This is the most definitive way to reduce the risk of exposure. For some individuals, the decision may be to avoid the surgery suite for a variety of different reasons. These may include difficulty standing during the length of surgical procedures, fatigue and exhaustion, severe nausea, fainting episodes, and safety concerns about animal size, position, and movement. Whatever the motivation for choosing to avoid anesthesia and/ or surgery during pregnancy, the decision deserves to be met by colleagues with validation and understanding.

BOX 2.1   Practice profile: Anesthesia and trying to conceive

Sara Wohlhueter-Page, DVM

When my husband and I decided to try for a child, we quickly determined we would need help. The at-home ovulation test kits never showed that I was ovulating. After seeing a fertility specialist and undergoing many tests, we were told there was a good chance that with medication and IUI we could become pregnant. We were thrilled. Then came the discussion about my career and how big of a risk doing surgery and anesthesia could be. While no clear right or wrong answer was given, after our consult with the specialist I knew what the right decision was for me. During veterinary school, I developed a strong love for anesthesia and surgery which blossomed even more once I was out in practice. However, I knew that if I continued as a surgeon, I would blame that and myself for every cycle with a negative pregnancy test. My husband, who works with me as a veterinarian as well, supported my decision and took on additional surgeries while I transitioned to seeing more appointments. My decision was hard but was absolutely the right one for me. For others in my position, their right choice is exactly the opposite.

PRECAUTIONS

For individuals who feel comfortable and choose to participate in anesthetic and surgical procedures during their pregnancies, there are several precautions that can and should be taken to help protect them and their babies. The American College of Veterinary Anesthesia and Analgesia (ACVA) recommends that these precautions be followed to protect *all* personnel, not just pregnant individuals.[3] Aside from complete avoidance, the recommendations to limit waste anesthesia gas exposure can be divided into three categories: engineering controls, administrative and work practice controls, and personal protective equipment (PPE).

ENGINEERING CONTROLS

When it comes to inhalant anesthetics, engineering controls correspond to maintaining the proper working order of the anesthesia machines and scavenging system. Anesthesia machines and all of their components must be checked often for leaks and repaired when needed. Personnel should check for leaks before every procedure, and the machine should be professionally checked and maintained at least yearly. The scavenger system must be checked regularly to make sure it is in good working order because an active scavenging system can reduce anesthetic waste gas in the air by up to 90%. Portable canisters can be used for passive scavenging. However, they are not quite as efficient and should be used only when traditional active scavenging methods are not available. Furthermore, the canister must be weighed with every use and discarded once it has gained 50 grams in weight.[3]

ADMINISTRATIVE AND WORK PRACTICE CONTROLS

Administrative and work practice controls encompass a number of best practices regarding administration of inhalant anesthesia to the patient. All of these are aimed at reducing the amount of waste anesthetic gas that escapes into the air where veterinary personnel will be exposed to it. The ACVA advises that mask and box (or chamber) inductions should be avoided.[3] These methods of induction involve administering large doses of anesthetic gas over a prolonged period of time and allow for significant amounts of anesthetic to leak out around the mask or from the induction chamber when it is opened. Avoiding these types of inductions also reduces the dose of inhalant anesthesia given to the patient, which is safer for them. Anesthesia should be induced by injection whenever possible, and the lowest necessary dose of inhalant anesthetic gas should be administered.[5]

The vaporizer should not be turned on until the patient is already connected and the endotracheal tube cuff is properly inflated.[3] Personnel should avoid sniffing any part of the anesthesia circuit to determine if the gas is flowing because this is an insensitive way to detect it and can result in exposure to large doses of anesthesia.[12] ACVA cautions against disconnecting the patient frequently and advises to only empty the rebreathing bag into the scavenging system. If the patient must be disconnected, the vaporizer should be turned off until the patient is reconnected.[3]

At the end of the procedure, the patient should be left connected to the anesthesia machine with the vaporizer turned off and oxygen on for as long as possible while the patient recovers. This will flush as much of the anesthetic out through the scavenger as possible.[3] One of the periods of highest risk of exposure to anesthesia is during recovery after the patient has been extubated.[1] Pregnant personnel should avoid exposure to the patient during these times if possible. They should not spend time in a recovery room with recently anesthetized patients because the gas concentrations in the air will likely be elevated. Prolonged contact with the rubber components of the circuit (like the rebreathing bag) that absorb and then slowly release gases should be avoided.[12] Pregnant workers should not participate in refilling the vaporizer or be present when this is done. A key-fill adapter can be used by non-pregnant personnel to prevent spills of anesthetic during refilling.[5]

Although not routinely practiced in veterinary medicine, there are ways to measure the air concentration of waste anesthetic gas to ensure that it is within acceptable levels and to measure an individual employee's exposure. Air monitoring programs can be designed with the guidance of an accredited industrial hygienist. Anesthetic dosimetry badges, much like radiology badges, can be used to monitor an individual's anesthesia exposure, but each dosimeter is generally only used for up to eight hours. Both of these monitoring tools have been recommended by NIOSH for use in veterinary practices to ensure worker safety.[3]

BOX 2.2 Practice profile: Our roles do not define us

Alyssa Mages, BS, CVT

There are many defining roles for me in my life, and two of the most important are that of mom and veterinary technician. One wouldn't initially think that these two would be in conflict given the nurturing aspects of both, but unfortunately that is not always the case for many in this profession.

I chose to become a mother, and, while difficult at times, it is the best decision I have ever made. My eldest is a fiercely independent and determined individual, and has been since her unexpected, but much celebrated, arrival. My youngest is steadfastly loyal and unwaveringly kind, all while being an intensely empathetic person. Not opposite, but not exactly complementary, yet always that familial bond remains strong.

Prior to my impending motherhood, I had been heading to veterinary school when we found out we were expecting. If I was a braver person I would have gone, but with my partner still in graduate school in Vancouver, BC, where we were living, my parents in Pennsylvania, and my friends scattered everywhere but western MA, I elected to not attend. Luckily I was a surgical assistant in a specialty/referral veterinary practice there, so I was still in my happy place!

It was my intention to work up until my due date or until it became physically impossible to perform my duties. That was my team's hope as well, and they were extremely supportive in order to keep me in the OR as long as possible. Given that the risk of anesthetic exposure decreases exponentially based upon the distance from the vaporizer and the patient, I felt more than comfortable continuing to assist in any/all orthopedic, abdominal, and neurosurgical procedures. I did avoid thoracotomies and was not present at induction or transfer of the patients upon my supervisor's insistence. A respirator was available for my use should I require or request it, but as my physical status shifted from stable to uncertain rather quickly – not quite preeclampsia, but hypertension and peripheral edema were present – I ended up transitioning to a non-surgical role for the remainder of my pregnancy. Following the birth of my daughter, I was very fortunate to be able to remain at home for up to 51 weeks of [partially] paid parental leave. While it may not have been my full salary, it was enough in conjunction with my partner's graduate student stipend for us to get by. I would have happily returned to work at this practice, but when my partner graduated from his master's program, it was required that he and I and our ten-month-old daughter head back to the US.

Upon our return to the States in 2008, I found a position as an ER assistant and on-the-job-trained my way up into more of a technician's role. I have my BS in marine biology, so I had a decent foundation in A&P and biochemistry, but it was rapidly becoming apparent I needed more, and not just academically. I was dealing with suppressed post-partum depression (PPD) and needed to step away and step into a place of healing. I left this practice, moved to the city where I worked as the lead technician for veterinarian who was also a good friend, and I enrolled at Manor College, all in the fall of 2009. My partner, being unable to find work, was happily ensconced in the role of stay-at-home dad, which gave me the peace of heart and mind to embark on this journey.

In May of 2012 as graduation was approaching, I became pregnant for the second time and made the difficult decision to leave my friend's practice in order to move out of the city and into a better school district for my daughter, as well as to expand my skills/knowledge at a larger ER/specialty practice in the suburbs of Philadelphia. At my initial interview I did not disclose the fact that I was three months pregnant, as I was concerned I would not be hired if that fact was known. It wasn't until I had been there for nearly a month that I let my manager know that I was in fact expecting. While the leadership team wasn't overly thrilled, they were incredibly supportive, ensuring I was relieved from the overnight shift into a mid-shift ER/ICU position. Upon my son's birth, there were gifts, flowers, and lots of well wishes. But there was also a lack of financial support and a very short maternity leave. I was only initially going to be able to spend six weeks at home – with a newborn and a kindergartener. I fought for, and was given, 12 weeks which, while still not truly enough time (is there ever?), was better than nothing. Finances were tight, energy was low, and emotions were high, but we made it through.

Here I am, nine years later with two incredible young humans in my life and a career of which I'm proud. I am one of the lucky ones, in so many aspects of that word, and I am very aware of this and grateful for it. I worked for understanding and supportive organizations that enabled me to continue to work in various capacities while also providing the resources I needed in order to do so. While my first pregnancy, birthing experience, and post-partum journey was vastly different from my second, I learned a great deal about myself, my capabilities, and my professional outlook from both. I wish that here in the US we took care of new families – mothers, fathers, and siblings – as well as they do in other countries, and that impending motherhood wasn't something to hide but instead is a cause for celebration, no matter what. In the work that I do beyond the veterinary realm, this is a cause that I fight for and will continue to do so. I am not defined by my roles as a veterinary technician, or as a mom, but they are an integral part of who I am and I would not have it any other way.

PERSONAL PROTECTIVE EQUIPMENT

Finally, PPE can be used to reduce an individual's risk of waste anesthetic gas exposure. Surgical masks are not sufficient to provide any additional protection from inhalant anesthetic gases, but organic vapor respirators with activated charcoal filters can be used for this purpose.[3,12] These types of respirators are typically available from local hardware stores. They must fit well to provide adequate protection, and they do increase the work required to move air. This may exacerbate breathing challenges the pregnant individual may already be facing, especially later in pregnancy.[13]

TAKE-HOME POINTS

- All recent evidence suggests that, with proper precautions, veterinary personnel can safely continue to work with inhalant anesthesia during pregnancy if they feel comfortable doing so;
- When a scavenging system is functioning and used properly, the risk of exposure to high concentrations of waste anesthetic gases is greatly reduced;
- Proper use of anesthetic equipment and following best practices in the administration of anesthesia to patients can further reduce the risks to veterinary personnel while also allowing for better, safer patient care;
- Anesthesia avoidance and/or use of an organic vapor respirator are also valid options to reduce risks associated with waste anesthetic gas exposure;
- There is room for further improvement with the adoption of air monitoring programs and/or the use of anesthesia dosimetry badges.

REFERENCES

1. Gold CTK, Beran GW. Occupational hazards to pregnant veterinarians. *Iowa State Univ. Vet.* 1983;45(1): 55–60.
2. Shirangi A, Fritschi L, Holman CD. Maternal occupational exposures and risk of spontaneous abortion in veterinary practice. *Occup Environ Med.* 2008;65(11):719–725.
3. American College of Veterinary Anesthesia and Analgesia. Commentary and recommendations on control of waste anesthetic gases in the workplace. *JAVMA.* 1996;209(1):75–77. Accessed March 13, 2022. http://vasinc.net/ACVAguidelines.pdf
4. American Society of Anesthesiologists. Occupational disease among operating room personnel: A national study. Report of an Ad Hoc committee on the effect of trace anesthetics on the health of operating room personnel. *Anesthesiology* 1974;41: 321–340.
5. Hazard Evaluation System and Information Service. Isoflurane may harm veterinary worker health. March 2019. Accessed April 7, 2022. https://cvma-inline.net/wp-content/uploads/2019/06/AnesthesiaGas-IsofluraneSafety.pdf
6. Johnson JA, Buchan RM, Reif JS. Effect of waste anesthetic gas and vapor exposure on reproductive outcome in veterinary personnel. *Am Ind Hyg Assoc J.* 1987;48(1):62–66.
7. Schenker MB, Samuels SJ, Green RS, Wiggins P. Adverse reproductive outcomes among female veterinarians. *Am J Epidemiol.* 1990;132(1):96–106.
8. Shuhaiber S, Einarson A, Radde IC, Sarkar M, Koren G. A prospective-controlled study of pregnant veterinary staff exposed to inhaled anesthetics and x-rays. *Int J Occup Med Environ Health.* 2002;15(4):363–373.
9. Zacher R. Maternal wellbeing and pregnancy outcomes in anaesthetic trainees. *Anaesth Intensive Care.* 2019;47(4):326–333.
10. American College of Obstetrics and Gynecology. Nonobstetric surgery during pregnancy. April 2019. Accessed March 13, 2022. https://www.acog.org/clinical/clinical-guidance/committee-opinion/articles/2019/04/nonobstetric-surgery-during-pregnancy

11. Levy R, Ing C. Neurotoxic effects of anesthetics on the developing brain. February 23, 2022. Accessed April 12, 2022. https://www.uptodate.com/contents/neurotoxic-effects-of-anesthetics-on-the-developing-brain
12. McLain Madsen L. Pregnancy in the workplace. Accessed March 13, 2022. https://www.vetfolio.com/learn/article/pregnancy-in-the-workplace
13. Centers for Disease Control and Prevention. Personal protective equipment—Reproductive health. October 28, 2019. Accessed February 19, 2023. https://www.cdc.gov/niosh/topics/repro/ppe.html

3

"You were exposed to what?"

Infectious diseases

During my last pregnancy, I was very careful to avoid handling cat fecal samples at work and scooping the litter box at home (although I did do this occasionally). But then one day, I apparently forgot about all my efforts to avoid toxoplasma exposure and I manually de-obstipated a cat. Of course, I was wearing gloves, and there were no noticeable negative consequences of that decision. I did laugh, however, when I later contemplated that the one time that I forgot to take precautions around cat feces was to perform a procedure that would result in maximum feline fecal exposure.

Pregnancy is a state of immunosuppression, which keeps the immune system from reacting to and rejecting the foreign life growing inside the body.[1] This immunosuppression has the secondary effect of reducing the body's natural defenses to fight off infectious diseases. Infectious disease exposure is a possible occupational hazard of veterinary practice, and some of these infectious diseases can threaten the health of pregnant individuals and increase the risk of adverse pregnancy outcomes. Sources of exposure can include direct animal exposure or exposure to animal surroundings, inhalation, contact with bodily fluids and excrements, fomite transmission, food and water contamination, auto-injection, and performing necropsies.

DOI: 10.1201/9781003406907-3

With proper precautions, the risk of complications related to infectious diseases during pregnancy can be greatly reduced in veterinary practice. However, one of the complicating factors in reducing the incidence of infectious disease exposure is failure to recognize the risk and use appropriate protective measures. In a 2022 survey of 469 large animal veterinarians, none of the respondents answered all the questions regarding zoonotic disease transmission correctly. This included questions about the risk and routes of zoonotic transmission of diseases that commonly cause abortion in livestock. Inconvenience surrounding transporting and wearing appropriate PPE in the field was listed as the top barrier to PPE use.[2]

While veterinary professionals are experts on the treatment and prevention of infectious diseases in their patients, they may sometimes underestimate their own risk of exposure. Given the increased vulnerability to infectious disease introduced by pregnancy, it is important for pregnant individuals and those who work with them to be well informed of their risks of exposure and how to mitigate these risks. This chapter presents some of the infectious diseases that may affect pregnancy and/or lactation that are most commonly associated with work in veterinary medicine. This list is not meant to be exhaustive, but the precautions listed here may help to protect against other infectious disease threats as well.

TOXOPLASMOSIS

Toxoplasma gondii oocysts are most prevalent in cats, sheep, horses, and pigs and can live in the environment for years.[3] Infected cats will shed oocysts in their feces for about two weeks after an initial infection. The oocysts take anywhere from one to five days to sporulate, which is when they are infective to humans. Handling cat feces can be a source of infection through fecal-oral transmission. There is also speculation that humans can become infected by inhaling sporulated oocysts from cat feces that are in dust kicked up by horses in an enclosed area. Other activities that can predispose to contact with *Toxoplasma* include meat inspection (one study showed that 92% of meat inspectors were seropositive) and performing necropsies. *Toxoplasma* can cause abortions in sheep, so examining sheep that have aborted and handling the abortus can increase risk of exposure.[3]

Contact with animals and animal feces is not the only source of exposure, however. About 50% of toxoplasmosis cases in humans come from eating infected meat such as lamb or pork that has not been sufficiently cooked because as much as 25% of meat from these animals can contain sporocysts. Other infections can originate from gardening without gloves (cysts can live in the soil for long periods of time), exposure to sand where cysts are present, or eating unwashed fruits and vegetables.[4]

Varying reports indicate the number of yearly human congenital toxoplasmosis infections in the United States to be anywhere from 400 to 4000.[4] Congenital toxoplasma infection can cause spontaneous abortion, premature birth, and/or a number of congenital problems such as encephalitis after birth, hydrocephalus at birth or shortly after, chorioretinitis, fever, skin eruptions, hepatomegaly, splenomegaly, convulsions, developmental disabilities, microcephalia, and deafness.[3] The most dangerous scenario is when the expectant mother is originally seronegative but contracts the infection during the second trimester of pregnancy. If they have already been exposed before the pregnancy and have immunity, the risk to a baby from a subsequent infection during the pregnancy is much lower.[3] About one-third of those who become infected with toxoplasmosis during their pregnancies will transmit the organism to the fetus through the placenta.[3]

PREVENTION

People who are pregnant or are planning to become pregnant should "adhere to strict hand hygiene" after gardening or touching soil and "avoid risky behaviors" such as eating undercooked meat, eating unwashed fruits or vegetables, eating raw shellfish, or consuming "unfiltered" water.[5] These are considered to be the most important precautions to take to prevent toxoplasma exposure and infection. Exposure to cats has not been shown to present as high a risk for toxoplasma infection as these other sources, likely because not all cats are infected, and those who are infected do not chronically shed oocysts. Furthermore, oocysts take one to five days to become infective. Scooping the litter box daily prevents human exposure to infected oocysts via this route.[5] Coupled with excellent hand hygiene, the risks of exposure and infection through contact with cat feces can be greatly reduced.[5] Individuals who have contracted a known acute toxoplasma infection (who are not immunosuppressed) are advised to wait one to three months before becoming pregnant.[5]

LISTERIOSIS

Sources of *Listeria monocytogenes* infection include sheep, goats, cattle, dogs (rarely), and fowl. The most common syndrome in infected animals is encephalitis. This can present as depression, self-isolation, fever, incoordination, torticolic/spasmodic contractions, paralysis of facial muscles and throat, ptyalism, circling, and head pressing. In the fulminant phase of infection, some animals may try to eat or make chewing movements. Some animals may suffer late-term abortions, and the fetus may be macerated.[3]

Other sources of *Listeria* infection include soil, plants, mud, pasture grasses, manure, silage (fermented grass, hay, or other fodder), and stream water.[3] *Listeria* can also be transmitted through unpasteurized milk and cheeses as well as from consuming ice cream, raw vegetables and fruits, sausages, hot dogs and lunch meats, and raw or smoked fish and other seafood.[6]

Pregnant individuals are ten times more likely than other people to become infected with *Listeria*, and those who are Hispanic are 24 times more likely to become infected (this is theorized to be from eating soft cheeses from unpasteurized milk).[7] *Listeria* infection can cause pregnancy septicemia, which can lead to stillbirth. Infected infants who survive may appear to be healthy at the time of birth but may later develop listerial septicemia or meningitis, both of which can be fatal.[3]

PREVENTION

Pregnant individuals should exercise caution when handling animals or secretions of animals that could be infected. This includes the appropriate use of gloves, face coverings, and protective clothing and washing hands carefully after any possible exposure. Pregnant personnel should also keep in mind that the animal's environment, including bedding and feed, can be sources of transmission.[8] Avoiding contact with any aborted fetal material during pregnancy is advised, and handling of raw milk from possibly infected animals should also be avoided.[3,8]

PSITTACOSIS

Chlamydophila psittaci (formerly *Chlamydia psittaci*) can be transmitted by more than 100 species of birds, especially cockatiels and parakeets. Infected birds may not show any signs, but if signs are present, they are typically respiratory in nature.[9] In humans, psittacosis causes flu-like signs, and fetal infection can cause fetal hepatitis, respiratory failure, and fetal death.[4] Sources of exposure include contact with birds or excrement.[4]

PREVENTION

Hand washing and PPE (including gloves and masks) are the mainstays of avoidance for veterinary professionals. Cage cleanliness, isolating infected birds, and avoiding overcrowding within cages are also very important.

Psittacosis can be difficult to diagnose because tests to detect it are not often immediately available and cases are rarely reported to the CDC. Therefore, it may not be in the purview of many providers. When diagnosed, it is treatable with antibiotics, and most people make a full recovery.[9]

BRUCELLOSIS

There are multiple *Brucella* species, each with their specific host species. These include cattle, goats, sheep, camels, dogs, pigs, wood rats, pinnipeds (seals, walruses, and sea lions), cetaceans (whales, dolphins, and porpoises), and the common vole.[10] In the United States, *Brucella* species have been eradicated from domestic livestock but can still be found in pet dogs and in wildlife reservoirs. International movement of animals can also introduce this organism into other animal populations.[11]

Transmission of *Brucella* species to humans can occur through contact with infected animals and/or their excretions, including from newborn animals, aborted fetuses, and bodily fluids associated with birth. The organism can also be inhaled, especially in slaughterhouses and meat packing plants. Laboratory employees working with this organism are also at risk for exposure.

Brucella vaccines ("Bangs" vaccine) for cattle, sheep, and goats (RB51, S19, and Rev-1) can lead to infection in animal health workers through needle sticks or exposure of eyes or wounds to the vaccine, which is a modified bacterium. There are some nonveterinary sources of exposure as well: hunting and exposure to animal wounds, uncooked or undercooked meat, and preparation of the animal carcass can also be associated with exposure to *Brucella* spp. Consuming raw and/or unpasteurized meat and dairy products from infected animals can also lead to human exposure.[10]

After human exposure, the *Brucella* organism undergoes hematogenous spread throughout the body, but it displays an affinity for the pregnant uterus and the placenta. There, the bacterium attacks placental cells and disrupts the maternal-fetal connection.[12] This can lead to miscarriage, and the risk is greatest during the first and second trimesters of pregnancy.[10,13] Neonatal brucellosis is rare and results from the transfer of the *Brucella* organism from mother to baby across the placenta, through bodily fluid exposure at the time of birth, or through breast milk. This can result in clinical signs such as fever, irritability, icterus, vomiting, trouble with feeding, and respiratory distress. Diagnostic workup of these cases may reveal hypotension, hyperbilirubinemia, thrombocytopenia, and pulmonary infiltrates. If the disease worsens, infants can develop hepatomegaly, splenomegaly, and lymphadenopathy. Sometimes the appearance of clinical signs of disease is delayed, or infants may remain asymptomatic.[13]

PREVENTION

Prevention involves using appropriate PPE, including rubber gloves, goggles, and gowns or aprons when handling animals, fluids, or tissue samples that could be infected.[13] All individuals are recommended to use full-length sleeves, eye protection, and an N95 or higher rated mask that is fit-tested when assisting with livestock births (e.g., farrowing sows). Similar precautions should be taken for assisting with cattle births if there is any potential exposure between the cattle and wildlife. The USDA also recommends testing all dogs for brucellosis before breeding. If this has not been done, rubber gloves, eye protection, and an N95 or higher rated mask should be used to assist with whelping.[14] Pregnant veterinary professionals should avoid examining animals who have had a spontaneous abortion or stillbirth, to pre-empt the risk of exposure to *Brucella* spp. They should avoid handling and administering the *Brucella* (or "Bangs") vaccines to cattle. If there is any exposure to *Brucella* during pregnancy, it is imperative to seek prompt postexposure prophylaxis and/or treatment in consultation with a healthcare provider.[13]

Q FEVER

Q fever is caused by the bacterium *Coxiella burnetii*. Common host species include goats, sheep, and cows, but the organism has also been isolated in wildlife, marine mammals, birds, reptiles, and domestic mammals, as well as 40 different species of ticks. Human transmission usually happens through inhalation of the organism in aerosolized soil or animal excrement. The organism can also be passed to humans through contact with feces, urine, milk, and birthing products from infected animals. Less commonly, it can be transmitted by consuming unpasteurized dairy products, through tick bites, or from human-to-human contact.[15]

In animals, most infections are asymptomatic. However, Q fever in ruminants can result in abortion, stillbirth, endometritis, mastitis, and infertility. Serologic testing is not a reliable way to determine if an animal is actively infected. Polymerase chain reaction testing of animal excretions or birthing products is much more reliable. Symptoms in humans are typically vague flulike symptoms, hepatitis, or pneumonia, and they present about two to three weeks after exposure, although many cases (up to 50%) are subclinical.

Adults are more likely to show clinical signs than children, and the overall mortality rate is low (<2%).[15]

In pregnant people who become infected during pregnancy or shortly before, there is a risk of miscarriage, stillbirth, premature birth, intrauterine growth restriction, or low birth weight. The mechanism for these complications is thought to be infection-related vasculitis or thrombosis (blood clots) that affect the health and effectiveness of the placenta. It has been shown that an infected mother can transmit the organism through the placenta to their baby, but no cases of birth defects have been reported. Some studies have detected C. burnetii in breast milk, but no infections in breastfeeding children have been reported through this route of transmission.[15]

The risk appears to be highest with acute infection during the first trimester. Untreated infection early in pregnancy can lead to miscarriage, and untreated infection later in pregnancy can lead to premature birth. Pregnant people are also more likely to develop chronic Q fever infections because of the failure of their immune systems to mount an adequate response. Furthermore, infection during one pregnancy can cause a risk of recrudescent infection during subsequent pregnancies.[15]

PREVENTION

This disease is difficult to prevent completely because many animal infections are asymptomatic, and diagnosis in animals requires specific testing. Furthermore, the organism can be aerosolized and spread through the air, therefore causing infections in some individuals with no contact with animals. Using appropriate PPE, including gloves, eye protection, and masks, can help reduce the risk.

Those who are pregnant or who could be pregnant should exercise caution in assisting with the delivery of animals and should avoid examining aborted fetuses in general. Antibiotic treatment is available and can reduce the potential risk for adverse outcomes for the baby from as high as 81% (if the disease is untreated) to 40% (when treated, and most of these cases do not involve fetal death). People concerned about adverse outcomes should discuss appropriate testing with their healthcare provider and seek infectious disease referral if needed.[15]

LYMPHOCYTIC CHORIOMENINGITIS

Lymphocytic choriomeningitis (LCM) is a viral disease of rodents. The primary host is the house mouse, and this species is also the main source of transmission of the disease to humans. Estimates indicate that about 5% of the house mouse population is infected with LCM and can shed it throughout their entire lives, and anywhere from 2%–5% of humans in urban areas have antibodies indicating exposure to the virus. Transmission typically occurs through exposure to the excrement or saliva from an infected animal or through a bite wound. Clinical signs typically occur 8–13 days after exposure and are usually manifested in two phases of fever and flulike signs. More severe cases can result in encephalitis, meningitis, hydrocephalus, and, possibly, myocarditis.[16]

Although mortality in humans is generally less than 1%, and severe clinical disease is not common, the organism can pass via the placenta from mother to baby. Infections within the first trimester can result in miscarriage, whereas infections later in pregnancy can result in birth defects including vision deficits, developmental disabilities, and hydrocephalus. This can also be true even if the mother showed mild or no clinical signs.[16]

PREVENTION

Prevention involves avoiding contact with wild mice. Pet rodents (including mice, hamsters, and guinea pigs) can also serve as a source of transmission, so extra precautions should be taken when handling these species during pregnancy. This includes using appropriate PPE, using good hand hygiene, preventing contact between pet rodents and wild rodents, and eliminating rodent infestations in homes. Treatment for this disease in humans is largely supportive, and it may include hospitalization and/or corticosteroids to reduce inflammation. This is another condition about which awareness in the human medical community is lacking, so advising health care providers of any possible exposure is essential.[16]

LEPTOSPIROSIS

Leptospirosis is a bacterial infection, having been described as one of the most prevalent zoonotic diseases in the world.[17] It is caused by one of several *Leptospira* species that affects more than 1 million people per year throughout the world. Although it is more common in tropical and subtropical climates, the disease can be found worldwide.[18] Leptospirosis can infect all mammals, and is shed in their urine, where it can contaminate water or soil and survive for weeks to months. Infected animals may have clinical signs of disease, or they may be asymptomatic.[19]

Human exposure occurs through direct contact with infected urine or other bodily fluids (other than saliva) or through water, soil, or food that has been contaminated with infected urine. The organism can enter the body through mucus membranes or through a break in the skin.[19] Most humans are exposed through water, such as during a flood or other natural disaster.[18] Veterinary team members may be exposed during their care of an infected patient.[19] Infection of a pregnant mother can result in passage of the organism through the placenta to the baby. This can result in miscarriage, fetal death, or death of the mother and/or newborn baby.[18]

PREVENTION

Prevention involves avoiding contact with infected urine, other body fluids, and water or soil that may be contaminated with infected urine. In veterinary practice, this may mean avoidance of a suspected or known case of leptospirosis by any pregnant team members. It should also mean using appropriate PPE, including gloves, eye protection, gowns, and rubber boots, if there will be any contact with potentially infected soil or water. Furthermore, antibiotic treatment is available if the condition is identified or suspected, so discussion with a healthcare provider is important.[19]

BOX 3.1 Practice profile: Leptospirosis exposure during pregnancy

Sara Sutherland, DVM

I got pregnant after a three- to four-year struggle with infertility—I didn't do IVF but was on a drug to make me ovulate, had a distressing miscarriage, and decided to try one more time and then give up. I had some scary bleeding early on, then had a pretty good pregnancy.

I wanted to keep working as long as possible and do as much as I could. In the first trimester I struggled a bit with nausea while pregnancy testing cows, but the second trimester was a breeze. Secretly I judged the vet in our clinic who was due at a similar time and refused to see any large animals or do any after-hours work because she was pregnant (even though she was still riding her horse over jumps).

I wasn't on call when the calving came in but it was a farm I was familiar with—lovely people—and I had been dealing with their recent calf abortion outbreak. We still didn't know why the abortions had happened, but we had ruled out BVD and suspected leptospirosis. I didn't think about lepto at all at that stage as I was busy working out who would go to the call. The vet who was on call rang me because he was on blood thinners several months after a heart attack so didn't want to take the risk of doing a calving—fair enough but it would have been nice of him to warn us in advance so we could arrange for another vet to cover situations like this! Another vet said he would go if absolutely necessary but would rather not as he was on call the following day and weekend so wanted a night off—also fair enough. I decided not to bother any of my other colleagues and just go myself.

The farm was about 40 minutes away from the clinic and out of cell phone reception. The heifer had been calving for a couple of hours. I double gloved as an extra precaution, put my hand in, and said, "this should be easy!"

Never, ever say that about a calving.

Four hours later I was rolling on the ground in a pool of blood and afterbirth, trying not to lie on my pregnant stomach, with cuts from embryotomy wire on my hands, gloves long gone, in the dark and rain. I was completely exhausted to the point of tears. I still wasn't thinking of the risk of lepto at that point. I couldn't get the last bit of wire around to split the hips no matter how hard I tried. In the end the owner elected to euthanize the heifer.

I rang my partner from the owner's house to tell him I was on my way home and to please walk the dogs. When I got home, I had a shower and went right to bed—too tired to eat but too hungry to sleep. I sent a message to my boss asking if I could start late the next day (her reply was "we don't actually expect you to do a four-hour calving at six months pregnant—you should have rung someone to come and help you!") The next day the lab results came back confirming leptospirosis in the aborting cows. Then I started thinking about leptospirosis.

Lepto is a very common disease in New Zealand—the most common occupationally acquired zoonosis. I knew several clients who had gotten very sick with lepto from their stock, some with long-term effects. One had to spend several weeks in ICU with multi-organ failure. Two recovered but couldn't drink alcohol even years later. I did what you should never do when you're worried about a disease and googled it. Then I learned that lepto can cause abortion in humans, even without the mother showing any signs. I booked a GP appointment with difficulty as I kept bursting into tears thinking about it—blame pregnancy hormones and stress from lack of sleep.

The GP said he had never heard of lepto but was going to ring the obstetrician. The obstetrician rang back and said he had spoken with an infectious disease expert in Wellington (capital city) who told him that lepto was uncommon in NZ. I informed him that he was mistaken, this disease is very common in NZ rural areas. He reluctantly prescribed a course of amoxicillin, to my great relief.

Baby was fine and is now a healthy three year old who loves cows and every other animal.

Lessons I learned: don't judge people for being cautious during pregnancy—every person had their own relationship with their pregnancy and their own tolerance for risk. Don't be afraid to get your colleagues to help with big or tricky jobs. Don't ignore the risk of zoonotic diseases, and don't be afraid to educate the human doctors if they know less about a disease than you do!

SWINE INFLUENZA

The H1N1 Influenza A virus originated in pigs before transforming into a human influenza virus that caused a pandemic in 2009. However, the human virus was not exactly the same as the virus circulating through pigs, and it was determined to be a mixture of multiple swine strains along with human and avian influenza strains. This reassortment produced a new viral strain that circulates only in humans. Although it is unlikely that humans can catch this particular strain of influenza from pigs or other animals, the possibility of future reassortment of influenza viruses exists.[20]

Pregnant people are at increased risk (about four to five times greater) of serious illness developing if they become infected with H1N1. During the 2009 pandemic, H1N1 posed the greatest risk to pregnancies by causing fevers. Fevers in the first trimester can contribute to neural tube defects and other congenital defects. Fevers during labor can increase the risk of seizures, encephalopathy, cerebral palsy, and death in newborn babies.[20]

PREVENTION

The CDC recommends that pregnant people and other high-risk individuals avoid direct contact with pigs that are known or suspected to be sick. When avoidance is not practical, prevention involves proper hand hygiene and appropriate PPE—goggles or face shields, N95 masks, gloves, and protective clothing. Because more pathogenic, zoonotic strains of influenza could emerge through reassortment without warning, pregnant individuals should exercise caution when examining any pigs with compromised respiratory signs. The seasonal human influenza vaccine likely does not provide direct protection against swine influenzas. Pregnant people should also avoid contact with sick people to minimize human-to-human transmission.[21] If any influenza infection is suspected, consultation with a health care provider is recommended, because supportive care and antiviral medication may be available.

AVIAN INFLUENZA

Zoonotic transmission of avian influenza is considered to be rare. An influenza A H5N1 strain was reported to have caused outbreaks and zoonotic transmission to humans for the first time between 2003 and 2005 in Indonesia, Thailand, and Vietnam.[22] Between December 2021 and March 2023, only ten human cases have been identified throughout the world, including one case in the United States.[23] Infection in pregnant people with this virus is also considered rare, but in the cases that have been reported in the 2003–2005 outbreak, four of six infected pregnant individuals died, and the two who survived had spontaneous abortions. One of those who died was reported to have slaughtered sick chickens from her backyard flock shortly before falling ill.[22]

PREVENTION

Good hand hygiene, minimizing contact with sick or dead birds (both wild and domestic), and wearing appropriate PPE (mask, goggles or face shield, and protective clothing) are recommended. As with swine flu, the seasonal flu vaccine is not expected to be protective against avian influenza. Antiviral medication and other supportive care, as noted before, can be provided to individuals suspected to be infected.[24]

COVID-19

Although COVID-19 is likely a zoonotic disease, the biggest concern with the SARS-CoV2 virus is transmission between humans. COVID-19 can be spread through the air from an infected person, by respiratory droplets landing on the eyes, nose, or mouth, or by touching surfaces that have previously been touched by an infected person and then touching the eyes, nose, or mouth.[25] Those who are pregnant or who have been pregnant in the last 42 days are at increased risk for severe illness (defined as needing hospitalization, intensive care, and, in some cases, a ventilator) compared with persons who are not pregnant.[26] This severe illness can lead to higher rates of preterm delivery, stillbirth, and possibly other complications.[26]

PREVENTION

The risk of exposure to SARS-CoV2 can be greatly reduced by avoiding contact with people who are already infected with the virus, washing hands often with soap and water (or using hand sanitizer if soap and water are not available), maintaining a 6-foot distance from others when possible, and wearing a well-fitting mask that covers the nose and mouth (ideally an N-95 or KN-95 mask).[5] Avoiding crowded areas and areas that are poorly ventilated can also reduce the risk.[5] The CDC and ACOG both recommend that pregnant people receive the COVID-19 vaccine.[5,20,27] At the time of this writing, the number of COVID-19 infections has been steadily dropping, and the more recent variants have proved to cause much less severe disease in the general population.

TAKE-HOME POINTS

- Zoonotic and other infectious diseases can pose a serious risk to the life of the pregnant individual and/or their baby;
- The risks with most of these diseases can be greatly reduced by using best practices in terms of appropriate PPE, performing good hand hygiene, avoiding contact with materials and tissues that are more likely to be infectious, and consulting a healthcare provider early and often with any concerns if there is a suspected exposure;
- Veterinary personnel will often have a much better appreciation for the exposure risk for certain zoonotic diseases compared with many human healthcare providers. Therefore, it is essential for pregnant individuals to advise their medical team of any suspected exposure or increased risk and to seek a specialist referral if indicated;
- Pregnant veterinary personnel should avoid contact with animals that have recently aborted a fetus, along with the fetus and any related fluids or tissue. In some cases, additional precautions are warranted when assisting with the delivery of veterinary patients;
- As seen with the COVID-19 pandemic and various influenza viruses, new diseases can emerge at any time and mutate, so maintaining best practices and vigilance is important.

REFERENCES

1. American College of Obstetricians and Gynecologists. Cytomegalovirus, parvovirus B19, varicella zoster, and toxoplasmosis in pregnancy. Accessed May 1, 2022. https://www.acog.org/clinical/clinical-guidance/practice-bulletin/articles/2015/06/cytomegalovirus-parvovirus-b19-varicella-zoster-and-toxoplasmosis-in-pregnancy
2. Cherry CC, Sureda MEN, Gibbins JD, Hale CR, Stapleton GS, Jones ES, Nichols MC. Large animal veterinarians' knowledge, attitudes, and practices regarding livestock abortion-associated zoonoses in the United States indicate potential occupational health risk. *J Am Vet Med Assoc.* 2022 Feb 16;260(7):780–788.
3. Gold CTK, Beran GW. Occupational hazards to pregnant veterinarians. *Iowa State Univ. Vet.* 1983;45(1): 55–60.
4. McLain Madsen L. Pregnancy in the workplace. Accessed January 15, 2022. https://www.vetfolio.com/learn/article/pregnancy-in-the-workplace
5. Petersen EE, Mandelbrot L, Weller PF. Toxoplasmosis and pregnancy. Accessed April 12, 2022. https://www.uptodate.com/contents/toxoplasmosis-and-pregnancy

6. US Food and Drug Administration. Listeria (Listeriosis). March 29, 2019. Accessed January 15, 2022. https://www.fda.gov/food/foodborne-pathogens/listeria-listeriosis

7. Centers for Disease Control and Prevention. Listeria (Listeriosis) people at risk—Pregnant women and Newborns. October 25, 2022. Accessed March 29, 2023. https://www.cdc.gov/listeria/risk-groups/pregnant-women.html

8. Dharma K, Karthik K, Tiwari R, Zubair Shabbir M, Barbuddhe S, Singh Malik SV, Singh RJ. Listeriosis in animals, its public health significance (food-borne zoonosis) and advances in diagnosis and control: A comprehensive review. *Vet Q.* 2015;35(4):211–235.

9. Centers for Disease Control and Prevention. Psittacosis. August 22, 2019. Accessed January 20, 2022. https://www.cdc.gov/pneumonia/atypical/psittacosis/index.html

10. Centers for Disease Control and Prevention. Brucellosis. November 2, 2021. Accessed February 2, 2022. https://www.cdc.gov/brucellosis/index.html

11. Pinn-Woodcock T, Frye E, Guarino C, Franklin-Guild R, Newman AP, Bennett J, Goodrich EL. A one-health review on brucellosis in the United States. *J Am Vet Med Assoc.* 2023 Mar 1;261(4):451–462.

12. Byndloss MX, Tsai AY, Walker GT, Miller CN, Young BM, English BC, Seyffert N, Kerrinnes T, de Jong MF, Atluri VL, Winter MG, Celli J, Tsolis RM. Brucella abortus infection of placental trophoblasts triggers endoplasmic reticulum stress-mediated cell death and fetal loss via type IV secretion system-dependent activation of CHOP. *mBio.* 2019 Jul 23;10(4):e01538-19.

13. Centers for Disease Control and Prevention. CDC National Center for Emerging and Zoonotic Infectious Diseases. Brucellosis reference guide: Exposures, testing and prevention. February 2017. Accessed January 20, 2022. https://www.cdc.gov/brucellosi/pdf/brucellosi-reference-guide.pdf

14. US Food and Drug Administration. APHIS Factsheet. August 2020. Brucellosis—What you need to know to protect yourself, your family, and your employees. Accessed January 20, 2022. https://www.aphis.usda.gov/publications/animal_health/fsc-brucellosis.pdf

15. Anderson A, Henk B, Fournier P-E, et al. Diagnosis and management of Q fever—United States, 2013 Recommendations from CDC and the Q fever working group. *Morb Mortal Wkly Rep.* 2013;62(3): 1–23.

16. Centers for Disease Control and Prevention. Lymphocytic choriomeningitis. May 6, 2014. Accessed February 2, 2022. https://www.cdc.gov/vhf/lcm/index.html

17. Rafique I, Khan F, Rabbani T, Ashraf SMK, Mahtab Md., Shukla I. Leptospirosis in pregnancy: Prevalence and outcome. *Indian J Obstet Gyn Res.* 2019:6(4):516–521.

18. Selvarajah S, Ran S, Roberts NW, Nair M. Leptospirosis in pregnancy: A systematic review. *PLoS Negl Trop Dis.* 2021;15(9):e0009747.

19. Centers for Disease Control and Prevention. Leptospirosis. Infection. June 9, 2015. Accessed January 30, 2022. https://www.cdc.gov/leptospirosis/infection/index.html

20. Carlson A, Thung SF, Norwitz ER. H1N1 influenza in pregnancy: What all obstetric care providers ought to know. *Rev Obstet Gynecol.* 2009;2(3):139–145.

21. Centers for Disease Control and Prevention. Take action to prevent the spread of flu between pigs and people. Accessed May 1, 2022. https://www.cdc.gov/flu/swineflu/prevention.html

22. Le TV, Phan LT, Ly KHK, et al. Fatal avian influenza A(H5N1) infection in a 36-week pregnant woman survived by her newborn in Sóc Trăng Province, Vietnam, 2012. Influenza Other Respir Viruses. 2019;13(3):292–297.

23. Centers for Disease Control and Prevention. Current U.S. bird flu situation in humans. March 6, 2023. Accessed March 29, 2023. https://www.cdc.gov/flu/avian-flu/inhumans.htm

24. Centers for Disease Control and Prevention. Prevention and antiviral of bird flu viruses in people. Accessed May 2, 2022. https://www.cdc.gov/flu/avianflu/prevention.htm#anchor_1647619154182

25. Centers for Disease Control and Prevention. How Covid-19 spreads. July 14, 2021. Accessed April 12, 2022. https://www.cdc.gov/coronavirus/2019-ncov/prevent-getting-sick/how-covid-spreads.html

26. Centers for Disease Control and Prevention. Pregnant and recently pregnant people at increased risk for severe illness from Covid-19. January 24, 2022. Accessed April 12, 2022. https://www.cdc.gov/coronavirus/2019-ncov/need-extra-precautions/pregnant-people.html

27. American College of Obstetricians and Gynecologists. Covid-19 vaccine during pregnancy is key to saving lives, medical experts urge. December 6, 2021. Accessed April 12, 2022. https://www.acog.org/news/news-releases/2021/12/covid-19-vaccination-during-pregnancy-is-key-to-saving-lives-statement

4

Drugs and chemicals

I will never forget the euthanasia appointment where I accidentally sprayed euthanasia solution all over my pregnant client. We had run out of Luer-lock syringes, and as I pushed the diluted euthanasia solution through the 12-mL syringe, the solution leaked (exploded!) out at the connection point with the needle, all over the bare legs of my grieving, pregnant client. I was embarrassed, but more importantly, concerned for her. Nothing bad happened, other than my feeling that I didn't come across very professionally to my client in her hour of greatest need. I worried that she may have felt that the health of her baby was threatened. We cleaned it up and I profusely apologized, and we proceeded with the euthanasia. I couldn't help thinking ... what if it had sprayed into her mouth or her eye, or into a skin wound? She probably still would have been fine with exposure of such a small magnitude. This experience changed the way I treated euthanasia solution in general, however, and the way I handled it during my own subsequent pregnancies.

Veterinarians may, by virtue of their profession, use drugs and other chemicals that are hazardous in general but that can be particularly hazardous to sperm or ova, a pregnant

DOI: 10.1201/9781003406907-4

person, and their baby. This includes drugs that kill cells, end a pregnancy, and even end a life. It is important to think about and be aware of the various drugs and other chemicals we may encounter daily so that they be handled safely and avoided when necessary. This chapter discusses the various drugs and other chemicals that can pose hazards to a developing fetus or embryo, or that can be transmitted to an infant by way of breast milk. It also explores ways to reduce the risk of harm to mother and baby.

ANTINEOPLASTIC DRUGS

Antineoplastic drugs (chemotherapy drugs) are used to treat cancer. They work by a variety of different mechanisms, but they are generally cytotoxic, especially to rapidly growing cells. Although this property allows them to specifically target cancer cells over other noncancerous cells in the body, it also places the fast-growing cells of the embryo and fetus at risk.[1]

Chemotherapy agents can be carcinogenic (causing cancer), mutagenic (causing mutations), and/or teratogenic (causing birth defects) to a developing fetus or embryo. Mechanisms of action of chemotherapeutic drugs include the disruption of nucleotide synthesis, division of cells, and/or transcription of DNA.[2] These changes may lead to spontaneous abortion or fetal death, congenital malformations, and/or low birth weight. Some of these drugs can also decrease fertility in individuals who are exposed to them. Additionally, some chemotherapy drugs can be passed in breast milk to infants; neutropenia has been noted in babies breastfed from mothers who were exposed to cyclophosphamide while lactating.[3]

The evidence of negative effects on pregnancies in personnel who handle chemotherapy drugs is mixed. Studies evaluating the risk of pregnant personnel to exposure, before the use of biologic safety cabinets and other protective measures were common, showed an association between occupational chemotherapy drug exposure and birth defects in babies born to nurses who handled chemotherapy drugs. They also showed a statistically significant correlation between occupational cytotoxic drug exposure in the first trimester of pregnancy and the rate of spontaneous abortions in exposed personnel.[3] However, a retrospective study of Australian female veterinarians found no increased risk of birth defects among the children of veterinarians who graduated after 1980, when many of the modern chemotherapy safe handling measures were instituted. This same study noted that among these more recently graduated veterinarians, those who continued to handle cytotoxic drugs regularly before they knew they were pregnant had increased risks of birth defects in their children in comparison to those who were planning their pregnancies and cutting back on their handling of chemotherapy drugs.[4]

As with radiation and anesthesia, there are rare instances where chemotherapeutic treatment is determined to be in the best interest of a person even during their pregnancy. When possible, it is avoided in the first trimester when many cell types are not differentiated and many vulnerable tissues are still developing. During the second and third trimesters, however, chemotherapy treatment in pregnant people is considered to be "relatively safe," and the recommendation is typically to not withhold it from people simply because they are pregnant. Retrospective studies revealed a 14% rate of major fetal malformations when chemotherapy was used in the first trimester of pregnancy compared with a 3% rate when chemotherapy was instituted during the second or third trimester, the latter of which is not significantly different from the rate of fetal malformations that occur in pregnancies without any chemotherapy exposure.[5] This suggests that, as with chemotherapy treatment in pregnant people, any occupational exposure to anti-neoplastic agents during pregnancy carries the greatest risk of fetal harm in the first trimester.

The National Institute for Occupational Safety and Health (NIOSH) details the multiple opportunities that exist for occupation exposure to chemotherapy agents. They include reconstituting and/or drawing up liquid drugs, administering injections by any route or priming an IV line, handling uncoated tablets, crushing tablets, or compounding a powder into a liquid suspension. Other sources of exposure include contact with the patient, any of their excrement or bodily fluids, and bedding and any equipment used to treat the patient. There can also be exposure when cleaning up after patients, including disposing of packaging and other waste used during patient care, cleaning areas used for drug preparation and patient care, disposing of unused drugs, and removing personal protective equipment (PPE) after use.[6]

PREVENTION

Some pregnant veterinary personnel may choose to avoid any contact with chemotherapy drugs and patients receiving chemotherapy throughout their pregnancy—and possibly even when they are still trying to conceive. The Oncology Nursing Society and the American Society of Health-System Pharmacists recommend that employers offer alternative duties that do not involve any exposure to hazardous drugs, such as chemotherapy drugs, to personnel who have reported that they are pregnant to reduce the risk of a negative pregnancy outcome as much as possible.[7] Gilani and Giridharan recommend that pregnant personnel not work in areas with a high risk of chemotherapy exposure for the first 84 days of their pregnancy. After 84 days, pregnant workers should still follow all precautions to avoid exposure as they handle chemotherapeutic agents. Lactating workers are also advised to avoid areas with a high risk of chemotherapy exposure, because many chemotherapeutic drugs can be passed through breast milk.[3]

For individuals who continue to work with chemotherapeutic agents during pregnancy, there are precautions that can be taken to reduce the risk of exposure to mother and baby. NIOSH has published guidelines on appropriate preventive measures for all personnel who handle chemotherapy drugs and/or who work in facilities where chemotherapy is administered.[6] These recommendations are comprehensive and include:

- Proper labeling and storage of chemotherapy drugs in an area that is limited in access to only essential personnel;
- The use of PPE such as chemotherapy gloves, eye protection, and an impervious gown when unpacking chemotherapy drugs, to provide protection in the case that a damaged container or other cause for a leakage is discovered;
- Not storing chemotherapy drugs with other drugs, and not allowing any food or drink in areas where chemotherapy drugs are stored, prepared, or administered;
- Preparing any hazardous drugs in a ventilated biological safety cabinet, and using an impervious gown and double-gloving with chemotherapy-rated gloves;
- Using double gloves, an impervious gown, and eye protection while administering chemotherapy drugs to patients;

- Using needleless or closed systems, if available, to prevent aerosolization of the drug during administration;
- Not disconnecting the IV line from the fluid bag once chemotherapy drugs have been introduced;
- Using PPE during the collection and disposal of any waste material or excess drug after chemotherapy administration;
- Use of PPE by personnel who clean surfaces used for drug preparation and patient care and housing to avoid exposure; and
- Washing hands with soap and water after removing PPE.

HORMONES

PROSTAGLANDINS

Prostaglandins are generally used more frequently in large animal medicine than in small animal medicine but can be used in both settings. According to the Zoetis prescribing information, prostaglandin F2 alpha, commonly known as Dinoprost or Lutalyse, is used to treat and manage various reproductive states in multiple animal species. In cattle, it can be used to synchronize estrus cycles, to treat silent estrus and pyometra, and to abort pregnancies. In dogs, it can be used to terminate an accidental pregnancy (termed "mis-mated" pregnancy). It is used in pigs to induce labor, and to manipulate the estrus cycles of horses.[8]

Humans are very sensitive to the effects of prostaglandins. These effects include luteolytic activity (destroying the corpus luteum that helps maintain a pregnancy) and causing contraction of smooth muscle, including that of the uterus. These two effects can significantly increase the risk of abortion. Later in pregnancy, there is still a risk of induction of labor and resultant fetal death. Prostaglandin F2 alpha is readily absorbed through the skin.[9]

PREVENTION

Individuals who are or could be pregnant should avoid handling prostaglandins or use extreme caution if they must handle them. This should include impervious gloves, skin and face/eye protection, and washing the skin immediately with soap and water in the case of any accidental contact. They should also be very careful to avoid accidental auto-injection[8,9]

DIETHYLSTILBESTROL

Diethylstilbestrol (DES) is a synthetic hormone that mimics estrogen. It is no longer commercially made for humans or animals but is still sometimes prescribed through compounding pharmacies for the treatment of estrogen-responsive urinary incontinence in spayed female dogs.[10] DES was used for the prevention of miscarriage in humans until 1971, when the FDA published a warning about the risks of birth defects in babies whose mothers took DES during their pregnancies. These birth defects included reproductive tract abnormalities, higher rates of infertility, problems maintaining a pregnancy, and clear cell adenocarcinoma of the vagina and cervix. Fetal exposure to DES also increased the risk of prostate and testicular cancer later in life. Mothers who took DES were found to have a slightly higher incidence of breast cancer.[11]

PREVENTION

The Safety Data Sheet for DES states that pregnant people should not have any contact with DES. They should not handle or administer the medication. Any accidental skin exposure should be treated by washing the area for at least 15 minutes.[12]

MISOPROSTOL

Misoprostol is a synthetic prostaglandin E1 (brand names Cytotec or Arthrotec [both from Pfizer]) that is used off-label in dogs, cats, and horses primarily to treat or prevent gastrointestinal ulcers, especially those related to nonsteroidal anti-inflammatory drug (NSAID) use. It can also be used to treat diseases of the uterus, kidneys, and skin, and to terminate a pregnancy.[13] It is this last indication that poses a risk to pregnant people. In the Cytotec prescribing information, it is reported that although misoprostol is used in humans to treat and prevent NSAID-related gastric ulcers, it is also used specifically in pregnant people to induce abortion and labor because it stimulates uterine contractions. This document lists multiple warnings from the FDA, including a warning about the risk of abortion, premature labor and birth, and/or uterine rupture.[14]

PREVENTION

Pregnant personnel should not handle misoprostol tablets or administer them to animals.[13] In the Safety Data Sheet from Pfizer for misoprostol, all individuals are cautioned against breathing in the dust from broken tablets or capsules. Any accidental skin exposure should be treated by washing the exposed area with soap and water, and eye exposure warrants flushing eyes with water for 15 minutes.[15]

OXYTOCIN

Synthetic oxytocin is used in both humans and animals to stimulate uterine contractions during pregnancy and milk letdown. It is also sometimes used to aid in uterine involution after birth, to promote the passage of a retained placenta, and to treat postpartum hemorrhage.[16] Because of its potent effects to stimulate uterine contractions, this drug could cause a miscarriage or premature labor through accidental auto-injection.[16,17]

PREVENTION

The USDA warns that personnel should take "particular care" to avoid accidental self-injection during pregnancy.[16] Oxytocin may also be harmful if it is absorbed through the skin, so skin exposure should also be avoided. Because this drug can act as a respiratory, skin, and eye irritant, all personnel should consider wearing impervious gloves, using eye protection, and avoiding inhaling the powder form of the drug. Any accidental skin exposure should be treated with soap and water for 15 minutes.[17]

SOLENSIA

Solensia (frunevetmab injection; Zoetis) is a new FDA-approved injectable monoclonal antibody indicated for the treatment of arthritis pain in cats. According to prescribing information, it works by binding to nerve growth factor (NGF) and blocking its effects. Because NGF is essential in the development of a normal nervous system in mammals, there is concern that this drug could interfere with this development in a growing embryo or fetus and/or infant. The prescribing information states that Solensia may pass through the placenta and/or into breast milk. The manufacturer does not comment on the potential for absorption of the drug through the skin in the case of accidental spillage.[18]

PREVENTION

People who are pregnant, lactating, or trying to conceive should avoid accidental exposure to this drug. Warnings recommend using "extreme caution" to avoid accidental self-injection.[18] Although the cautionary statements stop short of recommending that pregnant individuals not handle or administer the drug at all, compete avoidance is the safest way to prevent accidental exposure.

SILEO

Sileo (dexmedetomidine oromucosal gel; Zoetis) is an FDA-approved treatment for noise aversion in dogs according to its prescribing information. It can be absorbed through the skin, eyes, and mouth of both animals and humans. In pregnant individuals, exposure could stimulate uterine contractions and/or lower fetal blood pressure.[19]

PREVENTION

In its prescribing information, the manufacturer states that it is meant to be administered to the oral mucosa by way of a preloaded multiple-dose syringe. This is often done by the dog's owner at home, but in the case that it is to be administered in the veterinary clinic, pregnant personnel should not handle this product at all per the manufacturer. Standard precautions for the handling of this medication by nonpregnant individuals include the use of impervious gloves.[19]

CLEVOR

According to the prescribing information, Clevor (ropinirole ophthalmic solution; Vetoquinol USA) is a dopamine agonist used for the induction of vomiting in dogs. It is administered into the conjunctival sac and readily absorbed into the bloodstream. Clevor has been shown to negatively affect embryo and/or fetal development in animal studies using rodents. Per the manufacturer, accidental human exposure can also result in "headache, nausea, vomiting, dizziness, orthostatic hypotension, and sleepiness."[20]

PREVENTION

The manufacturer warns that individuals who are pregnant, planning to become pregnant, or breastfeeding should avoid contact with this drug. Even personnel who are not pregnant or breastfeeding are advised to wear gloves and eye protection while handling this medication. Any accidental exposure should be treated by flushing the affected area and seeking medical attention.[20]

METHIMAZOLE

The prescribing information for Felimazole (Dechra), the formulation of methimazole that is FDA-approved for use in cats, describes it as a drug that inhibits the production of thyroid hormones. Because methimazole is known to readily cross the placenta and concentrate in the fetal thyroid gland in both animals and humans, it is listed as a human teratogen. Methimazole can also be passed in high concentrations to infants via breast milk, where it can interfere with the nursing baby's thyroid function.[21]

PREVENTION

The human warnings in the prescribing information for Felimazole state that people who are pregnant, who may become pregnant, or who are breastfeeding should wear gloves when handling this medication. They should also wear gloves when handling any excrement or bodily fluids from treated cats, including urine, vomit, and feces, as well as their litter. All individuals are recommended to not break or crush the Felimazole tablets and to wash their hands well after any exposure.[21]

SODIUM PENTOBARBITAL

According to the Safety Data Sheet, sodium pentobarbital, commonly known as euthanasia solution, can be lethal to humans with ingestion of 2–3 g of the drug. At a concentration of 390 mg/mL, this would equate to an ingestion of more than 5–7 mL of the solution, much more than any accidental splash that entered the mouth would likely be. The solution is also reported to cause irritation with skin or eye exposure and with inhalation, but no more serious effects via these routes of exposure. Both pentobarbitone (the free-acid form) and ethyl alcohol (another ingredient in the solution) have been shown to have fetotoxic effects, but only with very high doses. Studies on fetal exposure to barbiturates showed coagulation disorders and withdrawal symptoms in neonates whose mothers were treated with barbiturates during pregnancy. In infants whose mothers were treated with IV doses of 600–750 mg of a barbiturate before delivery, 40% of babies had "moderate to severe" respiratory depression, with delayed establishment of normal respiration. Ingestion of ethyl alcohol in pregnant people has been shown to cause effects similar to fetal alcohol syndrome, but only when the dose is sufficient to cause toxicity in the mother.[22]

PREVENTION

The safety precautions recommended by the manufacturer for the use of sodium pentobarbital are the same for pregnant and nonpregnant individuals. Rubber gloves and eye protection are recommended when using this product. No respiratory precautions are recommended other than working in a room with normal ventilation. Needles should not be recapped by hand to avoid accidental needle prick or injection, and the entire syringe and needle should be disposed of in a puncture-resistant container.[22] Use of a syringe with a Luer-lock connection can also help to prevent accidental exposure through leakage of the solution at the needle-syringe connection.

METHYL METHACRYLATE

Animal studies involving methyl methacrylate (MMA), a liquid used in the mixing of bone cement for orthopedic surgery procedures, have implicated it as the cause of birth defects in rodents. These birth defects included growth restriction, skeletal abnormalities, and an increased risk of fetal resorption. The results of these animal studies have fueled concern that a similar risk could exist for humans.[23]

In a survey of female orthopedic surgeons, 41.7% of respondents indicated they would leave the operating room during the mixing of bone cement if they were pregnant, and 24.7% would leave the room if they were lactating to avoid exposure. In addition, 8.4% of respondents indicated that concern over MMA exposure had influenced the particular subspecialty of surgery they had chosen.[24]

Occupational exposure to MMA is by inhalation of the chemical from the air during mixing. The Environmental Protection Agency permits occupational exposure up to 100 ppm in air averaged over an eight-hour period, and signs of acute toxicity typically are noted around 1000 ppm. These signs include irritation of the mucous membranes, eyes, and skin.[23]

Studies in pregnant and lactating orthopedic surgeons that have attempted to identify the risk of occupational exposure to their babies have failed to document any increased risk. One such study reported that the maximum air concentration of MMA during hip replacement surgery was found to be 280 ppm, well below the concentration needed for toxicity to be expected.[25] Inhaled air concentrations for surgeons and other personnel present in the operating room were measured in another study and found to be 100 ppm at the highest.[26] A comparison of serum and breast milk samples from lactating surgeons with some MMA exposure with those of controls without MMA exposure found no detectable MMA in either serum or breast milk.[27]

CHLORAMPHENICOL

Chloramphenicol is a broad-spectrum antibiotic used in dogs, cats, and horses to treat a variety of bacterial infections. This drug is rarely used systemically in humans because of a small but serious risk of bone marrow suppression and irreversible aplastic anemia in about 1 in 10,000 humans. Chloramphenicol is readily absorbed through the skin and can bc passed across the placenta and through breast milk.[28] Concerns about the toxicity of chloramphenicol in pregnancy seem to originate from studies that identified "gray baby" syndrome in neonates less than 28 days old who were given intravenous, intramuscular, or oral chloramphenicol for the treatment of life-threatening cases of meningitis. Gray baby syndrome was characterized by "vomiting, respiratory distress, abdominal distension, ashen gray cyanosis, flaccidity, and cardiovascular collapse." The study reports that the "serious" reactions, a few of which resulted in death, were only noted in babies who were prescribed higher than normal doses or who were overdosed.[29]

No definitive evidence could be found of congenital defects caused by chloramphenicol exposure during pregnancy. A study looked at topical chloramphenicol eye medication use in pregnant individuals and found no statistical difference in the rate of birth defects between those treated with topical chloramphenicol eye drops or ointment during pregnancy and those who were not exposed.[30] Another study compared birth defect rates between pregnancies with oral chloramphenicol exposure in the first trimester and those without any exposure and found no increased incidence of congenital defects.[31]

PREVENTION

Although the risk of congenital defects is minimal with chloramphenicol exposure, there is still an "idiosyncratic" risk of "irreversible marrow aplasia" that is not dose dependent and associated with a high rate of mortality for any individual who has exposure.[29] Thus, caution is still warranted. Occupational exposure to chloramphenicol can occur through handling the drug in any of its formulations or inhaling the powder form. All personnel are advised to use gloves and a mask when working with this medication and to wash their hands after any accidental exposure. PPE should be used when cleaning up any vomitus or excrement from animals taking this medication to avoid exposure via this route.[28]

FORMALDEHYDE, FORMALIN, AND GLUTARALDEHYDE

According to the Centers for Disease Control and Prevention (CDC) and Frazier, all three of these related compounds are used as disinfectants and as preservatives for organs and tissue samples, and for cadavers used in anatomy laboratories.[32,33] In addition to being a known carcinogen, formaldehyde is reported to have the potential to cause miscarriage, cancer, and/or fertility problems in animal studies. This compound can cross the placenta, and although most work exposures are not high enough to cause formaldehyde to enter breast milk, there is the rare possibility of this occurring. Formaldehyde can be absorbed through the skin or through mucous membrane exposure, and it can also be inhaled.[32]

Per the CDC, the Occupational Safety and Health Administration's permissible exposure limit for short-term exposure to formaldehyde in nonpregnant workers is 2 ppm, and the time-weighted average is 0.75 ppm over eight hours. NIOSH measured formaldehyde levels at one mortuary college to ranged from 0.5 to 6.1 ppm, with an average of 1.4 ppm.[32] Studies involving medical students who had formaldehyde exposure in their gross anatomy laboratory over a 15-month period did not reveal any increased level of chromosomal changes in peripheral blood lymphocytes. The average air levels of formaldehyde in this study were less than 1 ppm. Another study looked at personnel involved with autopsies in a hospital. The formaldehyde levels in their workplace were measured to range from 0.61 to 1.32 ppm. No increase in mutations was noted in urine samples from these workers. There were a few studies that showed a positive correlation between formalin or formaldehyde exposure and the risk of spontaneous abortion in laboratory workers and postal workers. No studies involving humans demonstrated decreased fertility associated with formaldehyde exposure.[33]

Glutaraldehyde exposure in the workplace has been associated with asthma in hospital workers at levels lower than the published threshold limit value. No studies, however, have identified associations between glutaraldehyde exposure at work and decreased fertility, or an increased risk of spontaneous abortions or birth defects. Not enough is known about the potential for its transfer through breast milk.[32]

PREVENTION

The CDC recommends avoiding exposure to formaldehyde and related compounds during pregnancy and breastfeeding. When exposure cannot be completely avoided, several precautions are recommended to prevent any possible adverse effects.

- Use appropriate PPE. This includes gloves made from butyl rubber, neoprene, or nitrile. Latex gloves are not sufficient, because formaldehyde and related compounds can break down the gloves within 15 minutes. Other recommended PPE include a respirator, eye protection, and an impervious apron or gown;
- Open windows, if possible, to improve ventilation and decrease the risk of exposure through inhalation;
- Wash skin and change clothing after working with formaldehyde or having any direct exposure;
- Use formaldehyde neutralization sheets to clean spills quickly;
- Rinse tissue samples with water before handling them to reduce exposure;
- Keep formalin/formaldehyde under a chemical fume hood;
- Remove contaminated waste in a sealed, labeled container to prevent accidental exposure;
- Use extra caution when opening the chest or abdomen during a dissection, because the formaldehyde exposure may be higher in these instances.[32]

NITROFURAZONE

Nitrofurazone topical ointment is an antibacterial ointment used to treat infections in dogs, cats, and horses. Its use is strictly prohibited in food animals, and it carries warnings for humans about a possible risk of carcinogenesis based on studies in rats where mammary and ovarian tumors were noted.[34,35] While some equine veterinarians report being advised to avoid handling it during pregnancy, the evidence to support this recommendation is scarce. The International Agency for Research on Cancer (IARC) reports that there is "inadequate evidence" of any cancer-causing effect of this drug in humans. Furthermore, the drug is "poorly absorbed" through the skin and mucous membranes in humans.[36] Some humans can have a hypersensitivity reaction to nitrofurazone.[35]

PREVENTION

Even though it is not considered to be easily absorbed through the skin, all individuals who utilize nitrofurazone should apply the ointment to the animal using a spatula. Gloves should be worn with any handling of this medication, and hands must be thoroughly washed afterward.[36]

ETHYLENE OXIDE

According to the FDA, ethylene oxide gas sterilization is an important way to sterilize certain medical devices that do not tolerate high temperatures including polymers, metals, and glass. It is also sometimes the only available option to sterilize devices with multiple layers of packaging or that have hard-to-reach components (like IV catheters).[37] Some veterinary clinics may use ethylene oxide sterilizers for some of their equipment.

OSHA reports that exposure to ethylene oxide in liquid form can result in "eye irritation and injury to the cornea, frostbite, and severe irritation and blistering of the skin upon prolonged or confined contact." Ingestion of the liquid can result in irritation of the stomach and hepatopathy. Inhalation of the gas form can result in "respiratory irritation and lung injury, headache, nausea, vomiting, diarrhea, shortness of breath, and cyanosis." Ethylene oxide exposure has been linked to cancer in animals and in humans, and to neurotoxicity and chromosomal damage in the exposed individual. Adverse reproductive effects have also been reported.[38] Animal studies measured increased rates of spontaneous abortion in laboratory animals experimentally exposed to ethylene oxide. Chromosomal damage to the sperm of male animals was also observed, resulting in decreased fertility.[39] A retrospective study evaluated ethylene oxide exposure in pregnant people working in the sterilizing units of hospitals in South Africa in 2004 and used a questionnaire to determine their pregnancy outcome. A significantly increased rate of spontaneous abortions was found in people who were considered to have been over-exposed to ethylene oxide during their pregnancies. Those with lower exposures did not have a statistically increased spontaneous abortion rate.[40]

PREVENTION

OSHA provides a fact sheet that details some of the most important precautions to take when working with ethylene oxide. They include observing the exposure limit of 1 ppm of ethylene oxide in the air as an eight-hour time-weighted average (eight-hour TWA). Permissible exposure limits (PEL) of 5 ppm are permitted for no more than 15 minutes. If an employer can document that the employee exposure does not exceed 0.5 ppm averaged over an eight-hour period, no additional precautions are required. However, if exposure may exceed this limit, employers are required to provide employees with information and training, conduct air level monitoring and employee medical exams, provide appropriate warning labels, and provide appropriate PPE to employees. The PPE indicated will depend on the risk of exposure. For any exposure that may result in contact with liquid ethylene oxide, goggles and skin protection are required. A respirator must be used if the air limits will exceed the 1 ppm TWA or the 5 ppm PEL, in an emergency, or when installing or servicing ethylene oxide sterilizers.[41] Given the reported effects on fertility in laboratory animals and increased spontaneous abortion rates in both animals and humans, pregnant individuals and all individuals who are trying to conceive may wish to avoid any ethylene oxide exposure.

PESTICIDES

According to the National Pesticide Information Center, there are multiple sources for pesticide exposure in veterinary medicine. These include parasiticides applied to animals, products applied to lawns and vegetation, baits intended for slugs, snails, insects, or rodents, and pesticides used or stored in barns, garages, or homes where animals may access them. Veterinary personnel can be exposed through direct handling of the animal, exposure to the environment, or contact with vomitus from an exposed patient.[42] The CDC reports that exposure to some pesticides during pregnancy can be associated with miscarriage, birth defects, or even learning or developmental disabilities in children who were exposed in-utero. Some pesticides are believed to pass into breastmilk.[43]

PREVENTION

The CDC indicates that safe exposure levels to pesticides during pregnancy or while breastfeeding have not been established. Therefore, pregnant and breastfeeding individuals should attempt to avoid contact with pesticides where possible. When this is not possible, or when examining or caring for a patient where pesticide exposure is possible or suspected, exposure can be reduced by using appropriate PPE. This may include goggles, protective clothing, and/or a respirator. It is important to follow all instructions on the preparation, dilution, and use of pesticides to prevent toxicity and to avoid entering treated environments for the time specified in the instructions. Changing clothes and shoes can prevent tracking pesticides out of the environment where they are being used.[43]

BOX 4.1   Practice profile: Fertilizer exposure during pregnancy

Alissa Kirchhoff, MS, DVM

On a foggy morning about six months into my first successful pregnancy, a pushy, not so great client met me as I unlocked my clinic front door carrying a "cat who could not breathe." Without much thought, I hauled her out of her carrier and onto my exam table. Before I could notice her rapid respiratory rate and overall depression, I was hit with a sharp, tangy smell. I quickly got her on oxygen and started asking questions.

My client went on to explain that he'd found this particular cat along a creek bank, where he'd passed his neighbor applying herbicide to an adjacent soybean field. The mention of herbicide combined with the funky odor caused my stomach to drop as I realized that in my rush, I'd done her initial exam gloveless. As I gloved up and bathed her to get any residue off, worst case scenarios ran rampant through my mind. I asked the owner if he happened to know what was sprayed, and to my dismay, he did not. A few quick calls later, my worst fears were confirmed. The field, and potentially the cat, were most likely sprayed with Paraquat, a broadleaf fertilizer that can cause respiratory depression if in contact with a compromised epithelium or mucosal membrane.

With the cat as stable as I could get her on a tight budget, I excused myself to drive to the emergency room with Poison Control on the line. Poison Control evaluated my initial risk, deemed that since I was breathing and talking my current risk was somewhat low, and instructed me to call my obstetrician for further instruction. Fortunately after consulting with colleagues, my doctor agreed that the baby and I would likely be just fine, but asked that I come in anyway to be safe. Thankfully this story has a happy ending, and I now have a feisty two-year-old daughter to show for it.

With a few years of perspective under my belt, I look back on that day and shudder. I no longer am as quick to tolerate pushy, disrespectful clients. I stand my ground, prioritize the health and safety of myself and my employees, and ask detailed history questions before rushing into an exam. In addition to holding a firm line, you will rarely see me sporting a baby belly and working with patients with gloveless hands or a maskless face. Zoonotic disease and toxin exposure don't always readily announce themselves, and it just isn't worth the risk. Ultimately, my clients are abundantly aware that I am a mother, daughter, friend, and wife first and a veterinarian second.

TAKE-HOME POINTS

- Advances in safety protocols and more routine use of PPE can greatly reduce the risk of negative pregnancy outcomes associated with occupational drug exposure;
- The fact that fewer exposures and/or negative outcomes have been reported with these advancements in safety only further underscores the importance of continuing to adhere to best practices to prevent or reduce exposure and safeguard the health and well-being of mother and baby;
- There is still room for improvement with PPE use in many situations;
- Certain drugs, including prostaglandins, Solensia, Clevor, and Sileo, should be avoided altogether during pregnancy and/or lactation;
- It is up to the pregnant worker in most cases to communicate the level of exposure to certain drugs with which they are comfortable during their pregnancy.

REFERENCES

1. Scheftel JM, Elchos BL, Rubin CS, Decker JA. Review of hazards to female reproductive health in veterinary practice. *J Am Vet Med Assoc.* 2017;250(8):862–872.

2. Weaver VM, Dailey VA. Pharmaceuticals. In: LM Frazier, M Hage, eds. *Reproductive Hazards of the Workplace*. New York: Van Nostrand Reinhold. 1998:289–311.

3. Gilani S, Giridharan S. Is it safe for pregnant health-care professionals to handle cytotoxic drugs? A review of the literature and recommendations. *E Cancer Medical Science*. 2014;8:418.

4. Shirangi A, Bower C, Holman DJ, Preen DB, Bruce N. A study of handling cytotoxic drugs and risk of birth defects in offspring of female veterinarians. *Int J Environ Res Public Health*. 2014;11(6):6216–6230.

5. Hepner A, Negrini D, Azeka Hase E, et al. Cancer during pregnancy: The oncologist overview. *World J Oncol*. 2019;10(1):28–34.

6. The National Institute for Occupational Safety and Health. Preventing occupational exposures to antineoplastic and other hazardous drugs in healthcare settings. Publication Number 2004-165. Accessed May 15, 2022. https://www.cdc.gov/niosh/docs/2004-165/pdfs/2004-165.pdf?id=10.26616/NIOSHPUB2004165

7. Wyant T. What is ONS's stance on handling chemotherapy while pregnant, breastfeeding, or trying to conceive? January 3, 2017. Accessed May 18, 2022. https://voice.ons.org/news-and-views/what-is-onss-stance-on-handling-chemotherapy-while-pregnant-breastfeeding-or-trying

8. Zoetis. Lutalyse information insert. March 2007. Accessed February 8, 2022. https://www.zoetisus.com/_locale-assets/dairy/products/factrel/pdf/lutalyse_fullpg_eng_lut12013.pdf

9. Gold CTK, Beran GW. Occupational hazards to pregnant veterinarians. *Iowa State Univ. Vet*. 1983;45(1): 55–60.

10. Forney B. Diethylstilbestrol (DES) for dogs. Accessed April 2, 2022. https://www.wedgewoodpharmacy.com/learning-center/medication-information-for-pet-and-horse-owners/diethylstilbestrol-des-for-dogs.html.

11. Schrager S, Potter BE. Diethylstilbestrol exposure. *Am Fam Physician*. 2004;69(10):2395–2400.

12. ThermoFisher Scientific. Diethylstilbestrol safety data sheet. January 19, 2018. Accessed May 17, 2020. https://www.fishersci.com/store/msds?partNumber=AC204130010&productDescription=DIETHYLSTILBESTROL+99%25+1GRDIE&vendorId=VN00032119&countryCode=US&language=en

13. Gollakner R. Misoprostol. Accessed February 8, 2022. https://vcacanada.com/know-your-pet/misoprostol

14. US Food and Drug Administration. Cytotec misoprostol tablets. November 2012. Accessed February 8, 2022. https://www.accessdata.fda.gov/drugsatfda_docs/label/2012/019268s047lbl.pdf

15. Pfizer. Misoprostol tablets safety data sheet. October 1, 2018. Accessed May 17, 2020. https://pfe-pfizercom-prod.s3.amazonaws.com/products/material_safety_data/misoprostol_tablets_1-Oct-2018.pdf

16. United States Department of Agriculture. Oxytocin: Livestock. October 12, 2005. Accesed March 1, 2023. https://www.ams.usda.gov/sites/default/files/media/Oxytocin%20TR%202005.pdf

17. Medisca. Oxytocin safety data sheet. Accessed March 1, 2023. https://www.medisca.com/NDC_SPECS/MUS/2492/MSDS/2492.pdf.

18. Zoetis. Solensia (frunevetmab injection) prescribing information. Accessed March 24, 2022. https://www2.zoetisus.com/content/_assets/docs/solensia-pi.pdf

19. Zoetis. Sileo (dexmedetomidine oromucosal gel) prescribing information. Accessed March 24, 2022. https://www2.zoetisus.com/content/_assets/docs/vmips/package-inserts/sileo.pdf

20. Vetoquinol. Clevor full prescribing information. Accessed April 1, 2022. https://www.vetoquinolusa.com/clevor-info

21. Dechra. Felimazole FDA prescribing information. Accessed May 3, 2022. https://www.drugs.com/pro/felimazole.html

22. Virbac. Euthazol material safety data sheet. Accessed May 4, 2022. https://northamerica.covetrus.com/Content/SDS/009444.pdf

23. Downes J, Rauk PN, VanHeest AE. Occupational hazards for pregnant or lactating women in the orthopaedic operating room. *J Am Acad Orthop Surg.* 2014;22:326–332.

24. Harper KD, Bratescu R, Dong D, Incavo SJ, Liberman SR. Perceptions of polymethyl methacrylate cement exposure among female orthopaedic surgeons. *J Am Acad Orthop Surg Glob Res Rev.* 2020;4(3):e19.00117.

25. McLaughlin RE, Reger SI, Barkalow JA, Allen MS, Dafazio CA. Methylmethacrylate: A study of teratogenicity and fetal toxocity of the vapor in the mouse. *J Bone Joint Surg Am.* 1978 Apr;60(3):355–358.

26. Darre E, Jørgensen LG, Vedel P, Jensen JS. Breathing zone concentrations of methylmethacrylate monomer during joint replacement operations. *Pharmacol Toxicol.* 1992 Sep;71(3 Pt 1):198–200.

27. Linehan CM, Gioe TJ. Serum and breast milk levels of methylmethacrylate following surgeon exposure during arthroplasty. *J Bone Joint Surg Am.* 2006 Sep;88(9):1957–1961.

28. Forney B. Chloramphenicol for dogs, cats and horses. Accessed May 6, 2022. https://www.wedgewoodpharmacy.com/learning-center/medication-information-for-pet-and-horse-owners/chloramphenicol-for-cats-and-horses.html

29. Mulhall A, de Louvois J, Hurley R. Chloramphenicol toxicity in neonates: Its incidence and prevention. *Br Med J (Clin Res Ed).* 1983 Nov 12;287(6403):1424–1427.

30. Thomseth V, Cejvanovic V, Jimenez-Solem E, Petersen KM, Poulsen HE, Andersen JT. Exposure to topical chloramphenicol during pregnancy and the risk of congenital malformations: A Danish nationwide cohort study. *Acta Ophthalmol.* 2015 Nov;93(7):651–653.

31. Czeizel AE, Rockenbauer M, Sørensen HT, Olsen J. A population-based case-control teratologic study of oral chloramphenicol treatment during pregnancy. *Eur J Epidemiol.* 2000 Apr;16(4):323–327.

32. The Centers for Disease Control and Prevention; The National Institute for Occupational Safety and Health. Formaldehyde—reproductive health. November 15, 2019. Accessed May 22, 2022. https://www.cdc.gov/niosh/topics/repro/formaldehyde.html

33. Frazier LM. Disinfectants. In: LM Frazier, M Hage, eds. *Reproductive Hazards of the Workplace.* New York:Van Nostrand Reinhold. 1998:257.

34. Drugs.com. Nitrofurazone ointment. Accessed August 3, 2022. https://www.drugs.com/vet/nitrofurazone-ointment.html

35. EquiMed. Nitrofurazone. June 17, 2014. Accessed August 10, 2022. https://equimed.com/drugs-and-medications/reference/nitrofurazone

36. International Agency for Research on Cancer. Nitrofural (Nitrofurazone). November 1997. Accessed August 3, 2022. https://inchem.org/documents/iarc/vol50/09-nitrofural.html

37. Food and Drug Administration. Sterilization for medical devices. August 3, 2022. Accessed August 10, 2022. https://www.fda.gov/medical-devices/general-hospital-devices-and-supplies/sterilization-medical-devices

38. Occupational Safety and Health Administration. Substance safety data sheet for ethylene oxide (non-mandatory). January 8, 1998. Accessed August 10, 2022. https://www.osha.gov/laws-regs/regulations/standardnumber/1910/1910.1047AppA

39. California Department of Health Services. Ethylene oxide fact sheet. March 1991. Accessed August 10, 2022. https://www.cdph.ca.gov/Programs/CCDPHP/DEODC/OHB/HESIS/CDPH%20Document%20Library/eto.pdf?TSPD_101_R0=087ed34 4cfab2000f0bbfd30264a26cccf3ee0257f204d859361ebd7a32648cbbfc49e480d6 b046208fa7bbd24143000a96a57471e51a1d9319a9c2a92fab6ed2faa57d5a71ac37 3d2d9af49da8c3a72a5f974789f16921a88865658f9969246

40. Gresie-Brusin DF, Kielkowski D, Baker A, Channa K, Rees D. Occupational exposure to ethylene oxide during pregnancy and association with adverse reproductive outcomes. *Int Arch Occup Environ Health.* 2007 Jul;80(7):559–565.

41. Occupational Safety and Health Administration. Ethylene oxide fact sheet. 2002. Accessed August 10, 2022. https://www.osha.gov/sites/default/files/publications/ethylene-oxide-factsheet.pdf

42. National Pesticide Information Center. Pesticide information for veterinarians. January 20, 2021. Accessed August 10, 2022. http://npic.orst.edu/health/vet.html

43. Centers for Disease Control. Pesticides: Reproductive health. October 28, 2019. Accessed August 10, 2022. https://www.cdc.gov/niosh/topics/repro/pesticides.html

5

"Go ahead doc, I got her…"

Injuries and ergonomic hazards

When I was in the last trimester of my first pregnancy, I was accidentally stuck in the abdomen with a needle and syringe containing a feline rabies vaccine after the feline in question objected to the vaccine and sent the syringe flying in the air. Thankfully, I pulled it out quickly and was no worse for the wear. A few weeks later, another cat attacked me, admittedly because I made poor choices while trying to draw blood from him by myself. This cat attached himself to my face with his claws, and in my efforts to remove them, he started gnawing on my fingers. I sought treatment right away for the bite wounds. My OB/Gyn asked me which antibiotic to put me on because she figured I would know more than she did on this subject. Even with treatment, the wounds swelled significantly and were very painful. I remember performing surgery and not being able to bend a couple of my fingers because they were so swollen. There was no negative effect on my baby, but that might not have been the case if I had not sought appropriate treatment for the bite wounds. These experiences reminded me of the importance of safe animal handling to prevent injury and how unpredictable animals can be.

Working with animals automatically introduces a degree of unpredictability and risk of bodily harm. The Bureau of Labor Statistics ranked veterinary services higher than police officers and fire fighters in terms of rate of reported occupational injuries in 2016. Forty-one percent of injury claims were for veterinary technicians; 23% were for veterinary

DOI: 10.1201/9781003406907-5

assistants; 27% were groomers, receptionists, kennel workers, and other staff; and 9% were veterinarians.[1] An estimated 50–67% of clinical veterinarians and 98% of veterinary technicians experience an animal-related injury during their careers.[2] Although not all injuries to a pregnant individual will result in harm to the embryo or fetus, there is at least the risk of harm. Apart from this, pregnancy can result in physical changes that make an individual more susceptible to injury. Any injury can add to the already great physical demands of pregnancy on the body. Furthermore, injury during pregnancy can result in missed days of work, pain and decreased mobility, financial loss, and increased emotional stress.

The most common types of occupational injuries in veterinary practice include bites, scratches, kicks, falls, and needlestick injuries. As in many other workplaces, the physical work involved in veterinary clinical practice can lead to back and/or limb strain, which can compound conditions that are more common during pregnancy such as foot and leg swelling, hypertension, carpal tunnel syndrome, and others. Some work settings include exposure to loud noises for an extended period, which can cause a risk of hearing loss to mother and baby. Other possible injuries in veterinary clinical practice include kicking, crushing, trampling, and cutting injuries. Mobile or ambulatory personnel face the risk of injury in an automobile accident while traveling to their patients. Rarely, contentious interactions with clients and others can result in physical assault. Although not all injuries are completely avoidable, there are precautions that veterinary personnel can take to reduce the risk. These include personal protective equipment (PPE), workplace controls, appropriate physical and chemical restraint, careful case selection, and proper use of motor vehicle seatbelts and air bags.

BITES AND SCRATCHES

Bites and scratches are a concern with almost any animal, but the vast majority sustained in veterinary practice are from companion animal species. In a study that surveyed veterinarians in Australia in 2006, 48% of respondents reported that in the previous 12 months they had sustained a work-related dog bite wound that broke the skin.[3] Likewise, 67% of respondents reported an occupational cat bite or scratch that broke the skin within the previous 12 months.[3]

The organisms most commonly cultured from animal bite wounds include *Pasteurella*, *Streptococcus*, *Staphylococcus*, *Bartonella*, and *Capnocytophaga* species. Even in a healthy individual, there is still some risk of serious complications from infection with any of these organisms.[4] There has also been a case of *Francisella tularensis* (tularemia) infection, which led to sepsis and death in a non-pregnant individual, following a cat bite.[5]

Pasteurella multocida is a common cause of infection after a bite or scratch from a dog or cat. It is reported to cause a rapid development of cellulitis, and it can progress to a chronic deep infection of the affected area and possibly osteomyelitis. Many cases can require surgical drainage and injectable antibiotics to successfully treat.[6] Cases of sepsis and endocarditis following *Pasteurella multocida* from a cat bite have also been reported.[7] *Streptococcus* spp. have been reported to cause cellulitis, endocarditis, necrosis, and sepsis, as the result of an animal bite.[8] *Staphylococcus* spp. have also been associated with cellulitis and lymphangitis, but not usually abscesses.[9]

Data documenting the risk of *Bartonella henselae* and other *Bartonella* species transmitted through a cat scratch to a pregnant person is scarce. In a very small study of eight pregnant women diagnosed with cat scratch fever, no negative consequences of the infection were found in either the pregnancies or the infants born to the infected individuals.[10.] A single case has been documented, however, in which a woman who had frequent contact with cats and had sustained scratches from a flea-infested cat had signs and symptoms consistent with cat scratch fever. The woman's twin children, one of whom died eight days after birth, were both diagnosed with *Bartonella henselae* and other *Bartonella* species, as was her partner. The researchers concluded that the mother likely was infected with *Bartonella* from a cat scratch for 30 years and then passed the infection through the placenta to both twins.[11]

Capnocytopahga infections can lead to heart attack, kidney failure, gangrene, and sepsis, all of which can be life-threatening.[12] The risks of *Capnocytophaga* infection during pregnancy also include chorioamnionitis, or inflammation of the membranes surrounding the fetus. This is most likely when the infection happens in the last few weeks of pregnancy. Additionally, *Capnocytophaga* infection increases the risk of low infant birth weight and neonatal sepsis.[13]

PREVENTION

Bite and scratch prevention efforts must be multimodal to be successful. Depending on the species and the circumstances, bite prevention may include equipment like a muzzle, a squeeze cage, a twitch, stocks, or chutes to position the animal appropriately. These tools may allow the veterinary team to have control of the animal's head and/or body, thus achieving safer physical restraint. They are not a substitute, however, for other safe and effective restraint methods, fear-free techniques to reduce animal stress, appropriate pain control, and sedation and anesthesia if needed. In some cases, PPE such as bite gloves can provide additional protection. Pregnant personnel should use extra caution to avoid bites or scratches and avoid patients and procedures that are higher risk for a bite or scratch to occur without appropriate chemical and/or physical restraint. Any bite or scratch wounds should be cleaned immediately, and the pregnant individual should contact their healthcare provider to determine appropriate treatment.

NEEDLESTICK INJURIES

Needlestick injuries are very common among veterinary professionals. Whereas in human medicine a needlestick can prompt immediate concerns about transmission of diseases such as the HIV virus or hepatitis, fewer such concerns exist in veterinary practice. This can lead to a more relaxed attitude toward safety around needles among veterinary professionals. In a 2012 survey study of Canadian veterinarians, 58% of respondents reported having suffered a needlestick injury within the previous five years. In another study, 74% of animal health technicians reported having at least one needlestick injury within the previous year.[14]

Needlestick injuries can result in the self-injection of hazardous drugs, exposure to infection, allergic reactions, and more. In a survey study of needlestick injuries in female veterinarians, the most common complication noted was localized pain and swelling that was characterized as mild. However, there were also reports of brucellosis infections, severe allergic reactions, and one pregnant veterinarian suffered a spontaneous abortion at 15 weeks after self-injecting a prostaglandin (Dinoprost tromethamine) through an accidental needlestick.[15]

PREVENTION

Multiple preventative measures are available to reduce the incidence of needlestick injuries in veterinary medicine. These include avoiding the recapping of needles when possible and using a single-handed scooping technique when recapping if a needle must be recapped. The hard plastic syringe casing in which some syringes come can be used to recap needles without having to hold the cap, but this does not completely prevent needlestick injuries. Needle recapping devices can also be used to recap the needle without risk of injury. A number of innovative safety devices are also available to prevent needlestick injuries. These include resheathable butterfly needles, bluntable needles, hinged recapping sheaths, and retractable needles. These all offer some variation of either covering or concealing the needle after use with less risk of injury. These devices have been shown to significantly reduce the rate of injury in human medicine.[16] They are often not used in veterinary medicine, however, due to their increased cost when compared to needles without these features.

Needles should always be placed in an approved sharps container, and no one should attempt to reach into a sharps container for any reason. Personnel should not walk around with an uncapped needle. Proper physical and/or chemical restraint is also important to prevent patient movement that could cause personnel to inadvertently stick themselves or a coworker while using a needle.[16] As referenced in the chapters on infectious diseases (Chapter 3) and drugs and chemicals (Chapter 4), pregnant personnel should avoid handling medications (prostaglandins and others) and vaccines (*Brucella* vaccine) that carry a high risk of harm from accidental self-injection.

FALLS

Pregnancy causes loosening of the joints and a shifting center of gravity caused by the weight of the baby and uterus. This can leave pregnant people more at risk for falling and serious injury as a result.[17] Veterinary practice and pregnancy aside, falls are generally reported to be the most common cause of workplace injury.[17] In pregnancy, falls are the second most common cause of trauma and hospitalization behind car accidents. Some estimates indicate that as many as 25–27% of pregnant individuals fall during their pregnancy.[18] In a 2003 retrospective study of work and non-work-related falls during pregnancy, most falls occurred between the fifth and seventh months of pregnancy, with the most likely causes being slippery floors, "moving at a hurried pace," and carrying an object or child while walking. Also of concern in work falls were the reports of the wearing of improper footwear (described as shoes that were "slick, loose, or backless" or that had a heel height of more than one inch).[19] Other factors that were statistically shown to be linked to an increased risk of falling at work were a loud work environment, having been employed for less than three months at the start of pregnancy, and working rotating shifts, all likely causes of stress, fatigue, and/or inexperience in their current role.[19]

The most common injuries attributed to a workplace fall were bruises, cuts, rolled ankle, sprain/strain, and broken bone(s).[19] In most falls, pregnant individuals who fall tend to shift their bodies in ways to protect their abdomen and the baby. This may make them more vulnerable to some of these other injuries. Apart from instinctual actions to protect the baby from injury, the mother's anatomy also helps to cushion any blow that the baby receives during a fall or other injury. Specifically, the pelvic bones, subcutaneous tissues, thick uterine wall, and amniotic fluid all help to keep the baby safe and unharmed in most falling injuries.[20]

PREVENTION

Using proper footwear is very important for safety in veterinary medicine in general, and even more important for preventing falls during pregnancy. This means avoiding shoes with an elevated heel and instead choosing shoes with proper support and thicker tread. Veterinary clinics can reduce the risk of falling by providing nonslip flooring and using warning signs to indicate any areas where the floor may be wet or uneven. Ambulatory veterinarians do not have control over the terrain in which their patients reside. It therefore becomes much more important to have proper footwear, to plan ahead, and to move carefully and slowly, with assistance when needed. Some environments may need to be avoided completely, especially in the later months of pregnancy. Avoiding carrying equipment or patients while walking is also important in preventing falls. Another major component of fall prevention in veterinary medicine is being aware of and reacting to movements of the animal, which can be sudden, unpredictable, and sometimes high-energy. Using appropriate physical and chemical restraint is imperative. Even minor injuries in a pregnant individual warrant assessment by a medical professional.[20]

NOISE

Veterinary professionals can encounter high noise levels in a variety of different settings due to the physical characteristics of their workplace and the animals present there. In slaughterhouses, noise is produced by the movement and vocalization of animals, the opening and closing of metal gates, shouting of workers, and the use of various instruments and machinery needed to move animals through the facility. A study of three bovine and swine slaughterhouses in Italy documented noise levels ranging from 54 dB to more than 100 dB (as measured through a smartphone app). The average noise levels ranged from 71 to 100 dB.[21] Animal shelters can also be high-noise environments due to multiple barking dogs and physical structures that cause reverberation of sound. A study of noise levels in animal shelters indicated that noise levels frequently exceeded 100 dB.[22] For comparison, NIOSH Recommended Exposure Limit (REL) for occupational noise is 85 dB as an eight-hour time-weighted average (TWA).[23]

Chronic occupational exposure to loud noise can cause hearing loss in the human fetus once the fetal auditory system is developed, which is somewhere after nine weeks' gestation. Although it was originally thought that the soft tissues and fluid surrounding a fetus significantly attenuated noise before it reached a fetus, later studies using a higher quality hydrophone demonstrated only a very small reduction in sound by the tissues and fluid around a fetus. These studies in sheep demonstrated fetal hearing loss with exposure to 120 dB of noise.[24] In humans, a blinded observational study of children aged four to ten years whose mothers reported occupational noise exposure during pregnancy noted a significant incidence of hearing loss in the children whose fetal noise exposure exceeded 85 dB as an eight-hour TWA throughout their pregnancies. No fetal hearing loss was attributed to exposures below 85 dB.[25] High noise workplaces were not found to increase the risks of premature delivery, low birthweight, or fetal death.[26]

PREVENTION

Although PPE can successfully protect the pregnant worker from noises loud enough to cause hearing loss, no such protection exists for a fetus. Any exposure exceeding the 85-dB limit requires PPE for the worker and may necessitate avoidance or reduced time of exposure for pregnant personnel to prevent fetal hearing loss.[27] This does not necessarily mean a pregnant individual will need to leave their job if they work in a high noise environment. Noise levels can easily be measured using smartphone apps and/or smart watches. This can allow personnel to make an informed decision about their risk. Ways to avoid consistent high noise level exposure can include limiting the time spent in higher noise areas, taking frequent breaks, and requesting reassignment to tasks with less high noise exposure during pregnancy.

KICKING, CRUSHING, TRAMPLING, AND CUTTING INJURIES

Kicking, crushing, and trampling injuries are most common in veterinary personnel who work with large animals such as horses, livestock, some wild animals, and zoo animals.[28] Working with large animals can cause injury from head butting or swinging of the head, being stuck between an animal and a solid structure such as a wall, fence, or chute, being stepped on, or being kicked.[29] Thirty-nine percent of respondents in one retrospective study of injuries in Canadian veterinarians reported having suffered a crush, kick, or trample injury, and another 7% of respondents reported having suffered a head injury.[14]

The effects of these types of injuries on a pregnant person and their pregnancy will vary based on the anatomic location and severity of the injury. Any injury to the head or vital organs of the mother can be life threatening to both the mother and the fetus. Severe trauma to the abdomen can result in hemorrhage, placental abruption, pelvic fractures, uterine rupture, fetal health complications, and fetal and/or maternal death. Less severe injuries can still result in pain, bruising, fractured bone(s), soft tissue strain, and other complications that can interfere with mobility and the ability to continue working through pregnancy.[30]

Cutting injuries in veterinary medicine are mostly associated with scalpel use. In a survey of Canadian veterinarians, 35% reported having sustained a scalpel injury at some point during the previous five years.[14] The severity of this type of injury can range from minimal to life-threatening or even fatal. The risk of accidental cutting can be increased with patient movement, attaching and detaching blade handles, and failure to promptly dispose of used blades.[31,32]

BOX 5.1   Practice profile: Injury in small animal emergency medicine

Amanda Modes, DVM

I was working overnights during my second pregnancy. I was at a small emergency clinic where the overnight shift consisted of myself and two technicians. Later in my pregnancy (maybe around 30 weeks?) I had a small scare that was due to not using my best judgment. We had a large breed dog (a great Pyrenees?) come in and we diagnosed a GDV. The plan was to stabilize and transfer him to a specialty facility. Unfortunately, the technicians were having trouble placing an IVC as he was not cooperating. For some reason, I decided to take charge by restraining the dog myself. He was positioned in sternal recumbency, and so I stood over him restraining for a cephalic catheter from above. He decided to buck and banged into my (not so small) belly. I felt fine at first, and baby was felt moving around, but later I noticed some discharge I had not had before. I then waited a few hours until the office opened and called my OBGYN. The doctor on call was thankfully not concerned since I felt otherwise fine and baby was still moving but offered to have me evaluated if I wanted. Thankfully everything turned out okay, and I went on to be induced at 41 weeks with a healthy baby. In hindsight I should have made several decisions that were better and safer for both the patient and myself.

Prevention

Having knowledgeable, trained assistance when handling animals can greatly reduce the risk of injury. In a retrospective study of risk factors for injuries in Minnesota veterinarians, those who always had the assistance of a technician had lower rates of injury than veterinarians who said that they did not always have a technician's help when working with animals.[33] The use of appropriate PPE is also essential. Helmets, Kevlar vests, steel-toed boots or shoes, and leather gloves should be considered to help protect personnel from serious injuries when treating large animals.[28,29] These types of PPE would not serve as a replacement for, but rather a complement to, comprehensive training on appropriate animal handling and restraint techniques.[28] This can include the use of halters; mechanical restraints such as chutes, stocks, or stanchions; and using chemical restraint when needed. Care should also be taken to avoid the pregnant individual being positioned between a large animal and a solid structure to avoid being trapped and/or crushed.[29]

Appropriate animal restraint techniques and equipment can help prevent most scalpel and cutting injuries as well. In most cases where a scalpel is used, the patient should be properly sedated or anesthetized, or at least adequately restrained with a local anesthetic in use at the site of the procedure. All these precautions serve to reduce patient movement that can result in an accidental cutting incident.[29] Other ways to reduce the risk include using scalpel blade removers, bringing portable sharps containers into the field when needed, and conducting training around the proper handling of scalpel blades. Establishing a practice of always placing a scalpel into a sharps container after using it and not leaving it for others can also help reduce the risk.[31]

BOX 5.2   Practice profile: Injury on an equine farm call

Katie Scott, DVM

I am a mixed animal veterinarian, and I work at a five-doctor practice where we share on call. When I found out I was pregnant, I moved all of my on-call assignments for the remainder of the year to the first half of the year (July 2021 baby). When I was 30 weeks pregnant, I went to a horse colic call and didn't take a technician with me. It was at an Amish horse training, but it was Sunday, so minimal hands were available. The Amish man that stayed back from church to meet me was not exactly helpful when it came to doing my physical exam, but he was able to hold the lead rope.

When I went to take the mare's temperature, she turned to bite at her abdomen and stepped on my foot while bumping into my belly. She just stood on my foot, and down I went. I had to scramble to get up and took a minute to call my sister, an OBGYN resident, who said I needed to get checked out. I was a little flustered as I was not certain of the damage, but I felt relatively fine. I called my boss and let him know I would need coverage until I got discharged.

I drove myself an hour to my hospital and checked in on the labor and delivery floor. My sister was luckily working, and she initiated all the tests. About 20 minutes in she came back in and told me that I would need to stay if I didn't stop contracting. I didn't feel any contractions at all, so I'm glad she had me go to the hospital. This relatively mild fall had me in the hospital for three days while they tried to stop my contractions (with magnesium) and prepare baby for an early exit (with steroids). I had to stay in the bed and couldn't walk around for 48 hours while on the magnesium, which was pretty rough. They finally got the contractions to stop, and they considered it safe for me to go back to work the next day.

For the remainder of my pregnancy my boss wouldn't let me see any farm calls, as he couldn't afford me to go on leave any sooner than necessary. I ended up working until I got a nasty GI bug and went into early labor. I ended up in the hospital again for three days. I had minimal change in my pregnancy status, so I thought I was just there for rehydration purposes. I sent my husband to get dinner, my sister went home for the night, and I was going to rest. I rolled over and my water broke. Forty-one minutes later my husband made it back to me and my sister ran in, got a gown and glove on, and helped deliver my little one! It was exhilarating!

MOTOR VEHICLE ACCIDENTS

According to one study, motor vehicle accidents (MVAs) cause 50% of traumatic injuries that occur during pregnancy and 82% of all trauma-related fetal deaths. This type of major trauma is more likely to result in placental abruption (where the placenta separates from the uterus), pelvic fractures, hemorrhage, and shock in the pregnant mother.[34] The fact that MVAs are a leading cause of fetal death has been attributed in part to the improper wearing of seatbelts or failing to wear one entirely by pregnant individuals, possibly because of a concern that the seatbelt would cause more harm to the fetus during an accident.[34] Failure to wear seatbelts in general and to follow safe driving practices can also increase the risk of serious injury. In a survey of a group of AVMA members (who were not necessarily pregnant), only 56% of respondents reported obeying the speed limit when driving to farm or house calls, and 15% admitted to not wearing seatbelts.[35]

PREVENTION

Pregnant people should continue to use their three-point seatbelt (shoulder and lap restraints) during pregnancy, with the lap belt placed low on the hips under the belly. Vehicle airbags should not be disabled. Even though there is the potential for injury from either the seatbelt or the airbag, the risk of injury is much higher without these safeguards.[17] It is also important to obey the speed limit and all traffic laws.

WORKPLACE VIOLENCE

Unfortunately, workplace violence is also a concern in veterinary medicine. Reports of violence against veterinary professionals can range from threats of violence to assault, rape, or murder.[36] The perpetrators are often angry clients whose pet has died, but current or former employees, burglars, and individuals in troubled relationships with employees have also been implicated. Violence and threats of violence may sometimes be directed specifically toward the veterinarian treating a patient, but no member of the veterinary team is excluded from this risk.[25] Although these types of assaults are still relatively rare, the effects have the potential to be devastating to any member of the veterinary team, whether pregnant or not.

PREVENTION

Just like many other types of injury or trauma, it is not always possible to completely prevent workplace violence. But there are steps that can be taken to reduce the risk in some circumstances, and this may correlate to lives saved. A client's behavior in the veterinary clinic is often a result of not only their experience with their pet in the clinic but also all their additional life stressors. Taking time to try and identify the cause of a client's negative reaction or behavior may help to de-escalate a confrontation. Avoiding being alone with an unknown or angry client can also be important. For ambulatory practitioners, this would mean not performing farm or house calls alone.[36]

In an article on veterinary workplace violence prevention, Smither interviewed Eugene A. Rugala, a former FBI agent and behavior profiler, for advice on preventing workplace violence. He recommends a "mindful, but not fearful" approach to violence prevention. As part of this, he identifies loss as a major trigger for potentially violent behavior in veterinary clients. Most of the time this refers to the death of an animal. When having to report the loss of a patient to an owner, Rugala recommends that veterinary staff use a calm tone, display empathy, and use active listening techniques to help the client feel acknowledged and validated. Although these techniques will be insufficient to prevent a violent reaction in some clients, for others it may help to prevent a negative interaction from escalating.[37]

Furthermore, Rugala points out that many violent incidents that occur have already ended by the time first responders arrive to assist. Therefore, it is important for employees to be "stakeholders in their own safety and security and develop a survival mindset comprised of awareness, preparation, and rehearsal." This includes instituting a policy against violent or threatening behavior of any kind and communicating it to staff and to clients. Rugala states it should also be on display in waiting areas and explained to new clients. Employees who have volatile relationships with a partner or former partner, particularly ones in which a restraining order is in place, should be encouraged to share this information with their employer so that the likelihood of a confrontation (or worse) can be reduced.[37]

Rugala also outlines recommendations for modification of the veterinary clinic building to increase security such as locked doors or glass partitions to protect the reception area, panic buttons in the examination rooms and waiting area, good quality lighting and security cameras, and having a code word to signal that the police should be called. He also recommends regular training of staff to recognize circumstances in which violent behavior might be more likely in a client or employee and to recognize abnormal behavior that could escalate to a confrontation or violence.[37]

ERGONOMIC HAZARDS

According to Nesbit, ergonomics refers to "the science of adapting job physical demands to the size, shape and capabilities of the worker."[38] Hales further defines ergonomic hazards as "physical stressors and environmental conditions that pose a risk of injury or illness."[39] For pregnant individuals, physical stressors can include any of the physical or physiologic changes caused by pregnancy, including pregnancy complications. When combined with the demands of the working environment, the risk of illness or injury can increase. These pregnancy complications can also result in missed work, lost paid time off, increased discomfort and pain, and decreased motivation.[40]

CHANGES IN THE BODY CAUSED BY PREGNANCY

The body undergoes a number of changes to accommodate a growing baby, and all of these changes can affect how a pregnant individual interacts with their work environment. Muscles and ligaments relax to allow the body to expand for the growth of the baby and delivery. This can create decreased joint stability. The curvature of the spine increases as the weight of the body shifts the center of gravity forward. Blood volume increases to support the growing baby, meaning more fluid must be dealt with by the cardiovascular system. At the same time, the weight of the uterus, baby, and fluids causes compression of the veins returning blood to the heart. Pregnancy also creates a state of hypercoagulability, increasing the risk of blood clots. The growing uterus prevents the diaphragm from expanding as fully, which can affect breathing. A pregnant person's shifting center of gravity can cause challenges with balance. The hormonal changes and increased workload on the body can cause significant fatigue.[38,41,42]

Apart from all the normal changes, some pregnant individuals also have added complications such as gestational hypertension (preeclampsia), gestational diabetes, incompetent cervix, or placenta previa, just to name a few.[38] What would normally be a lower risk work activity can exacerbate one or more of these conditions. Table 5.1 lists some of the most common environmental conditions that veterinary professionals may encounter in the workplace and the possible complications that may result in pregnant individuals.

While many of these environmental conditions are difficult to eliminate from veterinary practice, steps can be taken to reduce the risk of injury or exacerbation of an underlying pregnancy condition. Table 5.2 lists risk mitigation strategies for each of the environmental conditions listed in Table 5.1.

Table 5.1 Possible complications exacerbated by work environmental conditions in pregnancy[27,38,40,42-44]

Environmental condition	Possible complication
Heavy lifting, pushing, pulling, bending, or other forceful exertion (including restraining patients)	• Exacerbation of incompetent cervix • Exacerbation of gestational hypertension • Muscle fatigue and/or injury • Back pain
Awkward postures (including palpating large animals, squatting or kneeling, performing surgery)	• Muscle fatigue and/or injury • Back pain
Prolonged standing	• Dizziness, lightheadedness, or fainting • Exacerbation of gestational hypertension • Increased fatigue • Swelling of feet and/or legs • Varicose veins • Foot and leg cramps • Back pain • Spontaneous abortion (increased risk only when standing more than eight hours per day continuously) • Preterm birth (increased risk only when standing more than eight hours per day continuously) • Placental abruption (rare)

(Continued)

Table 5.1 (Continued)

Environmental condition	Possible complication
Prolonged sitting (including during airplane travel and driving)	• Swelling of the feet and/or legs • Increased risk of blood clot development, which can lead to pulmonary thromboembolism • Back pain
Repetitive upper extremity use (typing, writing, grooming, dental work, blood collection, performing surgery)	• Carpal tunnel syndrome (weakness, numbness, pain, and/or abnormal sensation in the hand from compression of the median nerve) • Ulnar nerve entrapment (numbness, abnormal sensation, and/or pain along ulnar aspect of hand and little finger) • Thoracic outlet syndrome (numbness/weakness of the arm down to the little finger from compression of the brachial plexus nerves and vessels between the clavicle and ribs) • Tendinitis/tenosynovitis (inflammation of a tendon or its sheath anywhere in the arm or hand)

Table 5.2 Risk mitigation of pregnancy complications at work[27,38,40,42-44]

Environmental condition	Risk mitigation
Heavy lifting, pushing, pulling, bending, or other forceful exertion	• The recommended weight limit for lifting during pregnancy will depend on the individual, what they are lifting, how far along in their pregnancy they are, and how often they are required to perform the activity. Many resources recommend not lifting anything over 25 pounds • Avoiding lifting heavy objects from the floor or overhead • Avoiding lifting or bending repeatedly (more than every 5 minutes or more than 20 times per day) • Bending the knees and keeping a wide stance when lifting is necessary • Avoiding walking around and/or climbing while carrying an object or patient • Using assistive devices (such as hydraulic lift tables) or help from others to lift objects or patients when possible • Stretching the back gently to help relieve discomfort • Strengthening back, abdominal muscles, and pelvic floor through regular exercise • Being careful not to twist the torso when lifting • Avoiding lifting objects that are difficult to grasp or hold onto • Use careful patient and procedure selection and appropriate physical and chemical restraint when needed

(Continued)

Table 5.2 (Continued)

Environmental condition	Risk mitigation
Awkward postures	<ul><li>Limiting the time and frequency spent in awkward positions</li><li>Taking frequent breaks</li><li>Taking time to gently stretch after being in one position for a long time</li><li>Using a stool or other assistive device to reduce awkward postures</li></ul>
Prolonged standing	<ul><li>Not standing for more than three hours at a time without a break to sit</li><li>Wearing compression socks to help with circulation and prevent swelling of feet and legs</li><li>Wearing a back brace if needed for lower back support</li><li>Staying hydrated and eating regularly to reduce the risk of fainting</li><li>Decreasing workload (hours worked and/or physical demands of work) if fatigue or other health problems persist</li><li>Wearing proper footwear that is supportive and comfortable, avoiding high heels.</li></ul>
Prolonged sitting	<ul><li>Not sitting for more than two hours without standing up</li><li>Propping the feet up when sitting if swelling is a concern</li><li>Using a chair that is appropriately sized and supportive that allows for the feet to rest on the floor (or a footrest), the whole thigh to be seated on the chair, and the back to be supported</li><li>Rotating the pelvis back when sitting (i.e. avoiding slouching) to prevent sacroiliac pain and using a lumbar support pillow</li><li>Supporting body weight with a hand on a desk or other surface when rising from a seated position</li><li>When sitting on the floor, adopting a cross legged position (i.e. tailor position) to have a low center of gravity and relaxed legs</li></ul>
Repetitive upper extremity use	<ul><li>Ensuring proper wrist position and support when typing or performing other tasks</li><li>Avoiding prolonged pressure on any part of the arm, wrist, or hand</li><li>Avoiding carrying objects in front of the body with the arms extended or resting objects on the shoulder while carrying them to prevent thoracic outlet syndrome</li></ul>

BOX 5.3 Practice profile: Safety and logistical tips for large animal/ equine work during pregnancy

Melanie Barham, DVM, PMP, MBA

CEO, VETS STAY GO DIVERSIFY

When I was pregnant with my daughter, I was in equine practice full time. The majority of my clients were sport horse people, so I did a lot of lameness work. As someone who was relatively short and small framed, my pregnant belly got in the way relatively early. I am short, so I have always had to adjust how I do things to make them work for my height, and pregnancy was no different. Here are my top tips!

If you don't have a technician, it is an absolutely critical time to hire one, or take on a student. The same goes for forming an alliance for an on-call group in your area.

Have a barn staff member jog for you, and have a technician do your flexions, particularly hind limbs.

Sitting hunched over on a stool for ultrasounds/shockwaves gets more challenging as pregnancy progresses. Sitting in a slightly reclined position with your equipment raised a bit higher can be really helpful.

Technicians are often underutilized in equine practice—make sure you are optimizing yours by having them do work that they are permitted to within your state/province. If it's been a challenge to implement before, this is a perfect excuse that clients absolutely understand; it could pave the way for a new, more efficient way of practice!

For some injections where you may not be as agile anymore (e.g., hind limb nerve blocks), have a technician hold the limb so you can approach from the other side of the horse and the technician can focus solely on restraining the limb. For hind limb joint injections, having the opposite limb lifted is also really helpful to give you just a little extra time to move away.

Re-evaluate your balance and abilities on a regular basis. With a pregnancy, your balance and agility are greatly reduced, and can suddenly change week to week. Things that felt fine in week 20 might all of a sudden make you off kilter in week 25.

As equine vets, we often put ourselves in dangerous positions without realizing it. Do a common-sense check before doing anything, and enlist the help of your technician or support staff to help with this too. It can be helpful to have a visual reminder like a sticky note on your vet box near your twitch or injection tray asking "is this the safest way to do this procedure?"

Ask a colleague either within your practice or a nearby practice to be your back up if you have to say no to a procedure.

Keep snacks and water well stocked in your truck in case of low blood sugar. Simple carbohydrates are really handy for morning sickness.

Compression socks can be really helpful to keep swelling down in your legs.

Try to take a more reasonable approach to hours, and consider booking shorter days with more desk time and catching up on records as pregnancy progresses. You may feel great, but things can change quickly.

Make a plan for client care early. Ensuring your clients know who to call before they need to, whether that's an associate or locum, or a neighboring practice, will help them feel at ease. Sometime a meet and greet is helpful for bigger barns. Knowing that you care enough to ensure they have continuous care from a trusted colleague means less likelihood of lost business and clients who know how much you care about them and their barns/horses.

Remember to go easy on yourself and take care of you.

TAKE-HOME POINTS

- Using appropriate methods of restraint and having an assistant to properly restrain patients can significantly reduce the risk of injury;
- Caution should be used in the handling of needles, knives, and scalpels to prevent injury;
- Most of the injuries described in this chapter typically do not have serious negative effects on the success of a pregnancy or the health of the baby;
- Even if the baby is unharmed, injuries themselves can create additional challenges for the pregnant team member;
- Careful selection of patients and procedures during pregnancy is recommended;
- Motor vehicle accidents represent a significant threat to the life and health of a pregnant mother and baby, especially if they cause blunt force trauma to the abdomen or pelvis. Seatbelts should always be worn properly during pregnancy, and all rules of the road followed carefully to avoid accidents and prevent injury;
- Sustained loud noise can represent a risk for fetal hearing loss. Caution should be taken to limit continued noise exposure above 85 dB during pregnancy;
- Workplace violence is a risk in veterinary medicine. Every workplace should have a plan to reduce the risk and act quickly if the need arises;
- The risks of ergonomic hazards can be minimized with careful planning and modification of work activities as needed.

REFERENCES

1. Cima G, Larkin M. Hurt at work: Injuries common in clinics, often from animals, and usually preventable. October 10, 2018. Accessed June 22, 2022. https://www.avma.org/javma-news/2018-11-01/hurt-work
2. Fowler H, Adams D, Bonauto D, Rabinowitz P. Work-related injuries to animal care workers, Washington 2007–2011. *Am J Ind Med*. 2016;59(3):236–244.

3. Fritschi L, Day L, Shirangi A, Robertson I, Lucas M, Vizard A. Injury in Australian veterinarians. *Occup Med (Lond).* 2006;56(3):199–203.

4. Maniscalco K, Edens MA. Animal bites. Accessed: May 30, 2022. https://www.ncbi.nlm.nih.gov/books/NBK430852/

5. Blackburn J, Tremblay E, Tsimiklis C, Thivierge B, Lavergne V. Overwhelming sepsis after a cat bite. *Can J Infect Dis Med Microbiol.* 2013 Summer;24(2):e31–e32.

6. Arons MS, Fernando L, Polayes IM. Pasteurella multocida–The major cause of hand infections following domestic animal bites. *J Hand Surg Am.* 1982 Jan;7(1):47–52.

7. Caserza L, Piatti G, Bonaventura A, Liberale L, Carbone F, Dallegri F, Ottonello L, Gustinetti G, Del Bono V, Montecucco F. Sepsis by *Pasteurella multocida* Ssin an elderly immunocompetent patient after a cat bite. *Case Rep Infect Dis.* 2017;2017:2527980.

8. Tré-Hardy M, Saussez T, Yombi JC, Rodriguez-Villalobos H. First case of a dog bite wound infection caused by Streptococcus minor in human. *New Microbes New Infect.* 2016 Aug 12;14:49–50.

9. Abrahamian FM, Goldstein EJ. Microbiology of animal bite wound infections. *Clin Microbiol Rev.* 2011 Apr;24(2):231–246.

10. Bilavsky E, Amit S, Avidor B, Ephros M, Giladi M. Cat scratch disease during pregnancy. *Obstet Gynecol.* 2012;119(3):640–644.

11. Breitschwerdt EB, Maggi RG, Farmer P, Mascarelli PE. Molecular evidence of perinatal transmission of Bartonella vinsonii subsp. berkhoffii and Bartonella henselae to a child. *J Clin Microbiol.* 2010 Jun;48(6):2289–2293.

12. Centers for Disease Control and Prevention. Signs and symptoms of *Capnocytophaga* infection. October 16, 2018. Accessed February 3, 2023. https://www.cdc.gov/capnocytophaga/signs-symptoms/index.html

13. Centers for Disease Control and Prevention. Risk of *Capnocytophaga* infection. October 16, 2018. Accessed February 5, 2023. https://www.cdc.gov/capnocytophaga/risk-infection/index.html

14. Epp T, Waldner C. Occupational health hazards in veterinary medicine: Physical, psychological, and chemical hazards. *Can Vet J.* 2012;53(2):151–157.

15. Wilkins JR 3rd, Bowman ME. Needlestick injuries among female veterinarians: Frequency, syringe contents and side-effects. *Occup Med (Lond).* 1997;47(8):451–457.

16. Weese JS, Jack DC. Needlestick injuries in veterinary medicine. *Can Vet J.* 2008;49(8):780–784.

17. American College of Obstetricians and Gynecologists. Employment considerations during pregnancy and the postpartum period. April 2018. Accessed May 27, 2022. https://www.acog.org/clinical/clinical-guidance/committee-opinion/articles/2018/04/employment-considerations-during-pregnancy-and-the-postpartum-period

18. Hrvatin I, Rugelj D. Risk factors for accidental falls during pregnancy – A systematic literature review. *J Matern Fetal Neonatal Med.* 2021;2021. 10.1080/14767058.2021.1935849

19. Dunning K, LeMasters G, Levin L, Bhattacharya A, Alterman T, Lordo K. Falls in workers during pregnancy: Risk factors, job hazards, and high risk occupations. *Am J Ind Med.* 2003;44:664–672.

20. The Mayo Clinic. Trauma in pregnancy: A unique challenge. October 6, 2017. Accessed June 22, 2022. https://www.mayoclinic.org/medical-professionals/trauma/news/trauma-in-pregnancy-a-unique-challenge/mac-20431356.

21. Iulietto MF, Sechi P, Gaudenzi CM, et al. Noise assessment in slaughterhouses by means of a smartphone app. *Ital J Food Saf.* 2018;7(2):7053.

22. Coppola CL, Enns RM, Grandin T. Noise in the animal shelter environment: Building design and the effects of daily noise exposure. *J Appl Anim Welf Sci.* 2006;9(1):1–7.

23. National Institute for Occupational Safety and Health; Centers for Disease Control and Prevention. Noise and hearing loss prevention. Accessed June 5, 2022. https://www.cdc.gov/niosh/topics/noise/

24. Griffiths SK, Pierson LL, Gerhardt KJ, Abrams RM, Peters AJ. Noise induced hearing loss in fetal sheep. *Hear Res.* 1994 Apr;74(1–2):221–230.

25. Lalande NM, Hétu R, Lambert J. Is occupational noise exposure during pregnancy a risk factor of damage to the auditory system of the fetus? *Am J Ind Med.* 1986;10(4):427–435.

26. Greenberg GN, Cohen BA, Frazier LM, DeHart RL. Noise, ultrasound, and vibration. In: LM Frazier, ML Hage, eds. *Reproductive Hazards of the Workplace.* New York: Van Nostrand Reinhold. 1998:401–418.

27. Nesbit, T. Ergonomic exposures. In *Reproductive Hazards of the Workplace.* New York: Van Nostrand Reinhold. 1998:431–462.

28. Lucas M, Day L, Shirangi A, Fritschi L. Significant injuries in Australian veterinarians and use of safety precautions. *Occup Med.* 2009;59(5):327–333.

29. Virginia Tech University Veterinarian and Animal Resources. SOP: Cattle restraint. Accessed June 19, 2022. https://ouv.vt.edu/content/dam/ouv_vt_edu/sops/large-animal/sop-bovine-restraint.pdf

30. Lavin JP Jr, Polsky SS. Abdominal trauma during pregnancy. *Clin Perinatol.* 1983 Jun;10(2):423–438.

31. Qlicksmart. Preventing the risk of scapel injuries to livestock handlers. Accessed April 5, 2023. https://www.qlicksmart.com/scalpel-injury-livestock-handlers/?v=7516fd43adaa

32. Love, J. Vet dies in accident. January 28, 2008. Accessed April 5, 2023. https://www.jhnewsandguide.com/news/top_stories/vet-dies-in-accident/article_4ad4a861-a6f9-58a3-8b61-68276c574fac.html

33. Gabel CL, Gerberich SG. Risk factors for injury among veterinarians. *Epidemiology.* 2002;13(1):80–86.

34. Krywko DM. Toy FK, Mahan ME, Kiel J. Pregnancy trauma. StatPearls [Internet] Treasure Island (FL): StatPearls Publishing; 2022 Jan. https://pubmed.ncbi.nlm.nih.gov/28613676/

35. Landercasper J, Cogbill TH, Strutt PJ, Landercasper BO. Trauma and the veterinarian. *J Trauma.* 1988;28(8):1255–1259.

36. Smither S. Workplace violence: A wake-up call for veterinary practices. *Veterinary Team Brief.* May 2015:51–54.

37. Smither S. Prevent workplace violence. *Veterinary Team Brief.* May 2015:57–58.

38. Hales, T. Ergonomic hazards and upper-extremity musculoskeletal disorders. In: PH Wald, GM Stave, eds. *Physical and Biological Hazards of the Workplace.* New York: Van Nostrand Reinhold, 1994:13–41.

39. Rogers B, Buckheit K, Ostendorf J. Ergonomics and nursing in hospital environments. *Workplace Health Saf.* 2013 Oct;61(10):429–439.
40. Francis F, Johnsunderraj SE, Divya KY, Raghavan D, Al-Furgani A, Bera LP,
Abraham A. Ergonomic stressors among pregnant healthcare workers: Impact on
pregnancy outcomes and recommended safety practices. *Sultan Qaboos Univ Med
J.* 2021 May;21(2):e172–e181.
41. Soma-Pillay P, Nelson-Piercy C, Tolppanen H, Mebazaa A. Physiological changes in
pregnancy. *Cardiovasc J Afr.* 2016 Mar–Apr;27(2):89–94.
42. Tapp LM. The pregnant worker and ergonomics ASSE professional development
conference and exposition 2001 January 1 American Society of Safety Engineers.
43. The University of British Columbia. Pregnancy and ergonomics. Accessed
February 3, 2023. https://hr.ubc.ca/health-and-wellbeing/ergonomics/pregnancy
-and-ergonomics
44. American Veterinary Medical Association. Veterinary ergonomic guidelines.
Accessed February 8, 2023. https://www.avma.org/resources-tools/avma-policies/
veterinary-ergonomic-guidelines

6

Stress and fatigue

As a type-A go-getter, I pretty much never included taking a mid-day nap in my agenda. There were so many things I could be accomplishing instead of sleeping! Then came pregnancy ... it literally knocked me off my feet! While most parents of newborns will attest that the old adage "sleep while the baby sleeps" rarely works out, sleeping before the baby comes is definitely a good idea. And since nighttime sleep is often elusive and interrupted, naps become a necessity for many pregnant people, even the always-busy, can't-seem-to-sit-still ones like me. There were many times when I was grateful my body basically forced me to shut down and spend some quality time with my bed and my pregnancy pillow.

There's no way around it—pregnancy is exhausting and stressful, even without a physically and mentally demanding job. The hormonal changes, the extra weight, the lack of sleep, and more can make just getting through the day a monumental task. Veterinary personnel in clinical practice also face many physical challenges that can affect their pregnancies. While we are navigating the challenges of veterinary practice, we are facing the stresses in our families and relationships, financial obstacles, physical and mental

DOI: 10.1201/9781003406907-6

health difficulties, and community and global struggles of many kinds. Managing the demands of both veterinary practice and pregnancy is no easy feat.

When reading this chapter, it is important to remember that most causes of stress and/ or fatigue will not be sufficient by themselves to cause an adverse pregnancy outcome. Many veterinary professionals navigate stressful jobs and have healthy pregnancies, even when they work long hours and irregular schedules. The goal of this chapter is to provide information, not to scare the reader. With a better understanding of the effects of stress and fatigue and the preventive steps and coping strategies to mitigate them, pregnant veterinary personnel can find ways to withstand the rigors of both their professional lives and pregnancy and hopefully have a good quality of life. At the same time, employers and colleagues can learn to become aware of the challenges faced by pregnant veterinary professionals and help them feel validated and supported.

STRESS

The *Oxford English Dictionary* lists multiple definitions for the word *stress*, to be applied to different contexts including psychology, physics, engineering, biology, and linguistics. One reads, "a state of mental or emotional strain or tension resulting from adverse or very demanding circumstances." Another reads, "physiological disturbance or damage caused to an organism by adverse circumstances."[1] These two definitions highlight both the mental and physical effects of stress on humans. Stress is pervasive throughout all ages, cultures, and environments. Stress can be associated with the everyday frustrations of routine life, major catastrophic events, and the chronic struggles and difficult decisions that fill the spaces in between.

During periods of stress, the sympathetic nervous system is activated: epinephrine is released into the bloodstream; heart rate, respiratory rate, and blood pressure increase; and blood is shunted away from internal organs to the skeletal muscles. The body is preparing to fight or flee, even if neither of these responses is appropriate for the particular stressor. Although thousands of years of evolution have made the need to fight off an attacker or run away from a wild animal less common, the human body is just as wired as ever to react in preparation for these events. Even though this physiologic preparation may keep us alive in physically threatening situations, it is also thought to contribute to significant distress when the threat is not as imminent.[2]

Anyone who has worked in veterinary medicine could easily create a list of common workplace stressors: client interactions, long hours, moral and ethical dilemmas, futile care, lack of owner compliance, poor patient outcomes, stressed or aggressive patients, toxic work environments, low pay, missed time with family, and the list goes on. Pregnancy stressors can include anxiety about the health of the baby, worries about finances, gender disappointment, pregnancy-related conditions that make everything more difficult, and many others. Veterinary professionals who already have one or more children can experience stress associated with caring for their children during pregnancy. There are too many other possible stressors to name. Everyone's experience with work stressors depends on various factors, including their coping skills, their support system, and other nonwork stressors.[2] Stress will never be eliminated from veterinary practice, pregnancy, or life in general. The good news is that most stressors will not be sufficient to cause a negative pregnancy outcome by themselves. Still, having a better understanding of how stress can affect fertility and pregnancy can help both pregnant veterinary professionals and those who work with them mitigate these effects.

EFFECTS OF STRESS ON FERTILITY

Various studies associate job stress with decreased male fertility. Stressful experiences like intense military training and medical residencies were associated with decreased testosterone levels.[3,4] Stress has also been shown to affect sperm quality. Studies in both animals and humans have demonstrated reduced sperm health during periods of stress as well.[5,6,7]

Some studies also suggest a correlation between stress and female infertility. In a prospective study of couples who were trying to conceive, increased salivary α-amylase levels, a marker of stress, were associated with a longer period of time before conception occurred and higher rates of infertility.[8] Another study measured both salivary and hair cortisol levels in people who were undergoing in vitro fertilization. They found that although increased salivary cortisol levels (an indicator of acute stress) were not correlated with a decreased likelihood of successful conception, increased hair cortisol levels (an indicator of chronic stress) were.[9]

EFFECTS OF STRESS ON PREGNANCY

Studies in animals and humans have looked at various stressors and their effects on early pregnancy. Pregnant rodents who were repeatedly subjected to known stressors all had higher rates of early pregnancy loss.[10] In humans, studies have found that people who miscarried were more likely than pregnant people who had not miscarried to have had a "severe life event" in the weeks to months before the miscarriage.[11,12] Later in pregnancy, studies documented an association between increased "life stresses" and adverse pregnancy outcomes such as low birth weight, premature delivery, transfer to a neonatal intensive care unit, and in some cases, neonatal death.[13,14]

Stressful jobs and long work hours were shown to be associated with preterm birth and other complications. A study of pregnant resident physicians found that the rate of preterm deliveries doubled in residents who reported working more than 100 hours per week in their first trimester.[15] Another study looked at pregnant active-duty military personnel and found that they had a higher rate of pregnancy complications than pregnant people married to active-duty military personnel.[16]

EFFECTS OF STRESS ON LACTATION AND BREASTFEEDING

Stress, including work-related stress, can have negative effects on lactation, milk production, and breastfeeding. Going back to work after maternity leave is one of the principal reasons reported for cessation of breastfeeding.[17] One study measured oxytocin levels in lactating people who were exposed to stressful tasks or noise and found oxytocin levels to be decreased when compared with the levels measured without the stressful stimuli.[18] Lower oxytocin levels in the bloodstream correlate with less complete milk letdown and possibly decreased milk production.[19]

STRESS SPECIFIC TO VETERINARY PRACTICE

Merck Animal Health recently completed their third Veterinary Wellbeing Study and published their results in early 2022. In 2021, a total of 2495 veterinarians and 448 staff members completed the survey. The main findings of the survey indicated that shortages of staff and veterinarians, along with stress, student debt, and suicide, were the most serious problems reported by participants. Other variables that were associated with increased burnout included working in a "chaotic" practice, having little control over how they do their work, working long hours, and working more evening and weekend hours. Among Hispanic, Black, and Asian respondents, however, the largest concern was a lack of diversity within the profession.[20]

Veterinarian well-being was decreased when compared with that in 2019, and veterinary staff well-being was found to be lower than that of veterinarians. Both groups had higher numbers of respondents reporting "serious psychological distress" when compared to 2019, and both indicated high rates of burnout. Among veterinarians, groups who reported disproportionately higher levels of distress included Hispanic and Black veterinarians, younger veterinarians, and small and mixed animal practitioners. While suicidal ideation among veterinarians was not found to be increased above 2019 levels, it had increased among veterinary staff members.[20]

MANAGEMENT OF STRESS IN VETERINARY PRACTICE

The Merck study identifies steps that can be taken both by the individual and by the employer to reduce work-related stress in veterinary team members. For the individual, having healthy coping methods is important. Suggested coping methods include:

- Engaging in regular exercise;
- Practicing breathing and relaxation techniques;
- Practicing mindfulness techniques;
- Maintaining good nutrition;
- Spending time with friends and family;
- Setting boundaries to maintain reasonable working hours;
- Working with a financial planner to help reduce debt and gain more financial control.[20]

For employers, the first step to improving employee well-being and reducing workplace stress is acknowledging the problem and encouraging employees to address it in their own lives. This includes providing information on access to mental health resources and creating a culture of support for taking time off work to address mental health challenges. Other important components include providing health insurance that covers mental health treatment and an Employee Assistance Program and helping employees understand which benefits are covered.[20]

In addition, employers should foster a healthy "work climate." The four elements of this, per the Merck study, are:

1. A strong sense of belonging to a team
2. A high degree of trust in the organization

3. Candid and open communications among team members
4. Sufficient time to provide high quality patient care.[20]

Many other recommendations can be added to this list, including:

- Providing proper staffing ratios;
- Paying living wages to support staff;
- Setting reasonable expectations for clients;
- Preventing and/or reacting to abusive behavior;
- Assuring psychological safety in the workplace;
- Adequate mentoring for newer graduates;
- Support of life–work balance and family needs;
- Exercising empathy and understanding when time must be taken off work for pregnancy complications, medical appointments, and sickness of parent and/or child(ren);
- Providing accommodations to work activities to protect the pregnant employee and adjust to their physical needs and limitations.

A pregnant individual who feels supported in these areas at work will be less stressed and more confident in continuing to work through their pregnancy and beyond.

BOX 6.1   Practice profile: Recovering perfectionist

Alicia Fierro, CVT

I've wanted to be a mom for as long as I can remember.

The qualities of motherhood seem like they should mirror veterinary medicine so well. Mothers are nurturers. We are protectors. We look out for those who need special care. So it's no surprise that mothers make wonderful additions to the veterinary team. But how do you fully embrace both roles without sacrificing anything?

As a lifelong perfectionist, I want to excel at everything I undertake. With that kind of mindset comes a certain amount of anxiety, at least for me. So when I found out that I was pregnant with my first child, my desire to overachieve kicked into gear. I felt an intense amount of self-inflicted responsibility to my child, and rightly so. I was going to do everything perfectly and give my baby the best start possible. But how do you do that when your job naturally involves a certain amount of risk? These were risks that I was willing to take as an adult, individual being. As a pregnant woman, however, I was going to need boundaries if I was going to protect my unborn daughter.

Ironically, I also feared that the pressure I felt to do everything "right" during pregnancy would negatively affect my baby. The stress of trying to be perfect could translate into consequences for her health, and I worried that I might harm her by trying too hard. In my ten years as a veterinary technician, I've known other women in the field who have felt the same way. Veterinary medicine tends to attract driven, high-achieving individuals who want to be the best at what they do, and who will be their own harshest critics if they fail. So it's no wonder that that translates into motherhood.

No decent, law-abiding veterinary hospital would ever ask a pregnant woman to expose herself to radiation, lift heavy boxes, or wrestle with aggressive dogs. But as a perfectionist, I struggled with feeling inadequate in my career. I hated asking for help. I didn't want to feel less able than my coworkers. Veterinary healthcare workers get things done—we are the superheroes! I questioned if I was truly contributing or "valuable" anymore.

But my baby was more important than my pride, so I asked for help. And to their immense credit, my coworkers stepped in and cared for me like family. Throughout both of my pregnancies, I've had the blessing of being supported by an incredible team who usually didn't even wait for me to ask—they proactively took burdens away from me and never had a single complaint. I still felt valued and like I was an important part of our hospital.

So support your pregnant team members and don't let them feel that they're not contributing. Ask for help when you need it. Realize that you bring more to the table than just your ability to trim nails or monitor anesthesia. As a recovering perfectionist, I'll forever be grateful for the times in my career when I got to see how truly wonderful the veterinary team is, and how we not only take care of our patients but each other.

FATIGUE

The *Oxford English Dictionary* defines fatigue as "extreme tiredness resulting from mental or physical exertion or illness."[21] Fatigue is often experienced in both veterinary medical practice and pregnancy and can therefore be compounded in pregnant veterinary personnel. A certain amount of fatigue is considered to be normal during pregnancy, especially during the first and third trimesters. For some people the fatigue persists throughout the entire pregnancy.[22]

The causes of fatigue in pregnancy vary by trimester. In the first trimester, progesterone surges can result in fatigue that can start as soon as the second week of pregnancy. The energy demands associated with the formation of the placenta and the increased blood volume a pregnant person needs are another cause of energy depletion.[22] Although the second trimester sometimes provides a break from the fatigue, some people in this stage of pregnancy will continue to be plagued by it as a result of interrupted sleep from restless legs syndrome and/or the need to urinate throughout the night. Fatigue in the third trimester can have many causes. These include the energy demands of the growing fetus, the exhaustion of carrying around the weight of the fetus and surrounding fluids, and the size and positioning of the fetus that can make it hard to get comfortable and exacerbate the need to urinate frequently.[22] Other possible causes of pregnancy fatigue include lower blood pressure; lower blood sugar; morning sickness; stress and anxiety; back, hip, or pelvic pain; and heartburn that can make it difficult to sleep.[23]

EFFECTS OF FATIGUE ON FERTILITY

Irregular schedules and changing work shifts (back and forth from night and day shifts, for example) have been linked to fertility difficulties in some studies. Irregular shift schedules in poultry slaughterhouses and canneries have been associated with irregular menstrual cycles in employees.[24] In another study, those who worked irregular shifts were almost twice as likely to take more than nine months to become pregnant than those who were not employed in shift work. There was no measured decrease in male fertility associated with shift work.[25]

EFFECTS OF WORKPLACE FATIGUE ON PREGNANCY

In addition to the discomfort and decreased quality of life for the pregnant person associated with work and pregnancy-related fatigue, there is evidence to suggest that some work conditions that cause fatigue may be associated with adverse pregnancy outcomes. A study of Danish pregnant healthcare workers suggested that workers who worked at least two night shifts per week were at increased risk of suffering a miscarriage after eight weeks of pregnancy. In addition, the risk of miscarriage appeared to increase with the number of night shifts worked between the 3rd and 21st weeks of pregnancy.[26] Another study found an increase in preterm births to be associated with "rapid cycling" scheduling, where shift schedules changed frequently.[27]

Working long hours has also been associated with an increased risk of adverse pregnancy outcomes. A study of Japanese physicians found that those who worked 71 or more hours per week had 3 times the risk of experiencing a threatened abortion and 4 times the risk of preterm birth when compared with those who worked 40 hours or less per week. Pregnant physicians who worked 51–70 hours per week had 2.5 times the increased risk of preterm birth when compared with those who worked 40 hours or less.[28]

MANAGEMENT OF FATIGUE IN PREGNANCY

Ways to mitigate fatigue include nutrition, exercise, and rest. In addition to consuming extra calories during pregnancy to help support fetal growth, pregnant individuals are recommended to look for foods that are rich in iron. Iron is particularly important in pregnant people as it is a vital component of hemoglobin in red blood cells that are needed to carry oxygen to all tissues and cells in the body. Just like in animals, anemia is associated with fatigue in humans, so keeping iron levels up can help to reduce this cause of fatigue.[22]

Ideally, pregnant people should also consume foods that are high in protein, as they can provide long-lasting energy and help combat fatigue (if morning sickness and food aversions allow). Eating small, frequent meals becomes increasingly important as pregnancy progresses because of the compression of the gastrointestinal tract by the growing fetus. It also has the added benefit of helping combat nausea and keeping blood sugar levels more stable, which can help fight fatigue. And, not surprisingly, drinking copious amounts of water during the day helps support all the body's functions and is an important part of minimizing fatigue during pregnancy.[22]

While it may sound counter-intuitive, exercise can help combat fatigue. The American College of Obstetricians and Gynecologists recommends 150 minutes of "moderate aerobic activity" per week throughout pregnancy unless a healthcare provider has instructed otherwise. The benefits of exercise include improved circulation (and therefore oxygen delivery to tissues) and decreased back pain and constipation, among others.[22]

Rest, including plenty of sleep, is very important during pregnancy. Ideally, pregnant people will spend at least eight hours in bed each night with at least seven hours of sleep. To improve the likelihood of having restful sleep, it can help to keep the bedroom "dark, clean and cold" by using blackout curtains and unplugging or blocking the light from any electronic devices, changing sheets regularly, and turning down the bedroom temperature. Naps can also be helpful to recharge, but late afternoon and evening naps should be avoided to prevent disruption of nighttime sleep schedules. Avoiding caffeine after lunchtime can also help both mom and baby relax. Exercise should also be scheduled earlier in the day so that it does not interfere with evening relaxation in preparation for sleep.[23]

MANAGEMENT OF FATIGUE IN VETERINARY PRACTICE

To help combat pregnancy fatigue, pregnant personnel should avoid working through breaks and/or skipping meals.[29,30] Those who sit frequently throughout the day should take frequent opportunities to stand and walk around. Likewise, those who are on their feet most of the day should sit and elevate their feet as much as possible.[29] Breaks to either stand or sit should happen about every 45 minutes throughout the workday.[30] These types of accommodations in veterinary medicine can mean not participating in a long procedure that requires standing for hours at a time, or at least having the option to sit for part of the time. Other adjustments can include sitting for examination of small animals and for conversations with clients and taking a walk (or a nap) at lunch time.

Those who work more than 40 hours per week or who have irregular shift schedules may wish to reduce their hours and/or modify their schedules during pregnancy. Ways to reduce the likelihood of staying late at work can include:

- Having enough time per appointment and a reasonable caseload per day;
- Finding ways to simplify record writing (templates, dictation software, a scribe);
- Utilizing email and texting programs to reduce time spent on the phone with clients;
- Accepting help from coworkers to stay caught up on work and avoid having to stay late.

TAKE-HOME POINTS

- Working during pregnancy is not associated with an increased risk of an adverse pregnancy outcomes;

- High levels of occupational stress may affect male and/or female fertility;
- High occupational stress levels early in pregnancy may increase the risk of spontaneous abortion in some people, and high stress levels later in pregnancy can increase the risk of some adverse birth outcomes;
- Practicing coping mechanisms to mitigate stress *and* having a supportive work environment can reduce the risk of adverse pregnancy outcomes and improve quality of life for the pregnant individual;
- Fatigue is a common part of pregnancy for many people, and it can be compounded by occupational demands;
- Working 40 hours per week or less is associated with a lower risk for threatened abortion and preterm birth;
- Working more than two night shifts per week or working irregular shift schedules may affect female fertility and increase the risk of an adverse pregnancy outcome;
- Making schedule changes and setting boundaries around working hours may help to reduce fatigue and allow for a lower risk of an adverse pregnancy outcome.

REFERENCES

1. Oxford University Press. Stress. Oxford English Dictionary. n.d. Accessed July 9, 2022. https://www.google.com/search?client=safari&rls=en&q=stress+definition&ie=UTF-8&oe=UTF-8
2. Gammon PJ, Dell D. 1998 Occupational and nonoccupational stress. In: LM Frazier, M Hage, eds. *Reproductive Hazards of the Workplace.* New York:Van Nostrand Reinhold. 1998:519–533.
3. Kreuz LE, Rose RM, Jennings JR. Suppression of plasma testosterone levels and psychological stress: A longitudinal study of young men in officer candidate school. *Arch Gen Psychiatry.* 1972;26:479–482.
4. Singer F, Zumoff B. Subnormal serum testosterone levels in male internal medicine residents. *Steroids.* 1992;57:86–89.
5. Cui KH. The effect of stress on semen reduction in the marmoset monkey (Callithrix jacchus). *Hum Reprod.* 1996;11:568–573.
6. Giblin PT, Poland ML, Moghissi KS, Ager JW, Olson JM. Effects of stress and characteristic adaptability on semen quality in healthy men. *Fertil Steril.* 1988;49:127–132.
7. Fukuda M, Fukuda K, Shimizu T, Yomura W, Shimizu S. Kobe earthquake and reduced sperm motility. *Hum Reprod.* 1996;11:1244–1246.
8. Lynch CD, Sundaram R, Maisog JM, Sweeney AM, Buck Louis GM. Preconception stress increases the risk of infertility: Results from a couple-based prospective cohort study–The LIFE study. *Hum Reprod.* 2014;29(5):1067–1075.
9. Massey AJ, Campbell BK, Raine-Fenning N, Pincott-Allen C, Perry J, Vedhara K. Relationship between hair and salivary cortisol and pregnancy in women undergoing IVF. *Psychoneuroendocrinology.* 2016;74:397–405.

10. Baines MG, Haddad EK, Pomerantz DK, Duclos AJ. Effects of sensory stimuli on the incidence of fetal resorption in a murine model of spontaneous abortion: The presence of an alien male and postimplantation embryo survival. *J Reprod Fertil.* 1994;102:221–228.

11. O'Hare T, Creed F. Life events and miscarriage. *Br J Psychiatry.* 1995;167:799–805.

12. Brandt LP, Nielson CV. Job stress and adverse outcome of pregnancy: A causal link or recall bias? *Am J Epidemiol.* 1992;135:302–311.

13. Williamson HA, LeFevre M, Hector M. Association between life stress and serious perinatal complications. *J Fam Pract.* 1989;29:489–494.

14. Hedegaard M, Henriksen TB, Sabroe S, Secher NJ. Psychological distress in pregnancy and preterm delivery. *BMJ.* 1993;307:234–239.

15. MaGann E, Nolan T. Pregnancy outcome in an active-duty population. *Obstet Gynecol.* 1991;78:391–393.

16. Klebanoff M, Shiono P, Carey J. The effect of physical activity during pregnancy on preterm delivery and birth weight. *Am J Obstet Gynecol.* 1990;163:1450–1456.

17. Gjerdingen DK, McGovern PM, Chaloner KM, Street HB. Women's postpartum maternity benefits and work experience. *Fam Med.* 1995;27:592–598.

18. Ueda T, Yokoyama Y, Irahara M, Aono T. Influence of psychological stress on suckling-induced pulsatile oxytocin release. *Obstet Gynecol.* 1994;84:259–262.

19. Dewey KG. Maternal and fetal stress are associated with impaired lactogenesis in humans. *J Nutr.* 2001;131(11):3012S–3015S.

20. Merck Animal Health. Veterinary mental health and wellbeing and how to improve them: Learnings from the Merck animal health veterinarian wellbeing study III. January 2022. Accessed June 9, 2022. https://www.merck-animal-health-usa.com/about-us/veterinary-wellbeing-study

21. Oxford University Press. Fatigue. Oxford English Dictionary. n.d. Accessed July 9, 2022. https://www.google.com/search?client=safari&rls=en&q=fatigue+definition&ie=UTF-8&oe=UTF-8

22. HealthParners. Fatigue during pregnancy: Causes, tips and what to expect. Accessed July 9, 2022. https://www.healthpartners.com/blog/fatigue-during-pregnancy/

23. Knisley K. Welcome to pregnancy fatigue: The most tired you have ever felt. February 25, 2020. Accessed July 19, 2022. https://www.healthline.com/health/pregnancy/pregnancy-fatigue

24. Messing K, Saurel-Cubizolles MJ, Bourgine M, Kaminski M. Menstrual-cycle characteristics and work conditions of workers in poultry slaughterhouses and canneries. *Scand J Work Environ Health.* 1992;18:302–309.

25. Bisanti L, Olsen J, Basso O, Thonneau P, Karmaus W. Shift work and subfecundity: A European multicenter study. European study group on infertility and subfecundity. *J Occup Environ Med.* 1996;38(4):352–358.

26. Begtrup LM, Specht IO, Hammer PEC, et al. Night work and miscarriage: A Danish nationwide register-based cohort study. *Occup Environ Med.* 2019;76:302–308.

27. Davari MH, Naghshineh E, Mostaghaci M, et al. Shift work effects and pregnancy outcome: A historical cohort study. *J Family Reprod Health.* 2018;12(2):84–88.

28. Takeuchi M, Rahman M, Ishiguro A, Nomura K. Long working hours and pregnancy complications: Women physicians survey in Japan. *BMC Pregnancy Childbirth*. 2014;14:245.

29. Kaiser Permanente. Tips for working parents-to-be. Accessed July 20, 2022. https://healthy.kaiserpermanente.org/health-wellness/maternity/second-trimester/at-work

30. Garcia E, McCrary-Ruiz-Esparza E. How to beat pregnancy exhaustion when you do paid work, too. December 31, 2019. Accessed July 20, 2022. https://www.inhersight.com/blog/insight-commentary/pregnancy-exhaustion%2042

Trying to conceive, fertility treatments, and pregnancy loss

When I was in my 20s, I could get pregnant without much effort. In my mid to late 30s, when I wanted to expand our family again, I found it to be more difficult (but not impossible), even when I was tracking my ovulation. I was not diagnosed with infertility, and I found out I was pregnant the very day I was supposed to go for a consult to discuss initial fertility treatments with my doctor. That pregnancy ended in a late miscarriage, an experience that changed my life and changed me forever. I went on to have two more successful pregnancies after that, with meticulous tracking and constant anxiety about the "what ifs." I have seen others take years to successfully conceive, some with extensive fertility treatment. And the more I talked about my pregnancy loss, the more I realized that I was not alone in that experience either.

Conception—it's how all humans begin, and how our species survives. Yet it is something that we have relatively little control over. For some, it happens very easily, or even by accident. For others, it can be challenging, or even impossible. Modern reproductive

DOI: 10.1201/9781003406907-7

technology has made it possible for many individuals and couples to have biological children who previously could not. But it is still not an exact science. This chapter reviews some of the current recommendations for maximizing the chances of unassisted conception as well as considerations around the timing of conception with relation to age and professional development. Also included is a discussion of fertility treatments and adoption. A review of pregnancy loss, grief and recovery, pregnancy after loss, and supporting grieving parents follows. Pregnancy termination and ways to support parents who pursue it are also included.

TRYING TO CONCEIVE

The experience of trying to conceive can be very different from one individual to another, and from one pregnancy to another. Although specific recommendations should come from a healthcare provider, there are some general recommendations that may help increase the likelihood of conceiving without assistance. According to the Mayo Clinic,[1] they include:

- **Tracking and predicting ovulation.** This can be done by monitoring cervical mucus consistency and/or body temperature. Over-the-counter ovulation kits can help determine the window of ovulation by measuring hormones in the urine;
- **Having sex regularly.** This can mean having sex every day or every other day near the time of ovulation, starting at the end of menstruation;
- **Maintaining a healthy body weight.** Being overweight or underweight can contribute to ovulation disorders;
- **Taking a vitamin supplement that contains folic acid.** Starting at least a few months before trying to conceive can reduce the risk of neural tube defects in an embryo;
- **Not smoking.** Tobacco negatively affects health and fertility;
- **Reducing alcohol consumption.** "Heavy" alcohol use may negatively affect fertility, and the Mayo Clinic recommends avoiding it entirely, if possible, when trying to conceive;
- **Reducing caffeine consumption.** This means consuming less than 200 mg per day, equivalent to one to two 6- to 8-oz cups of coffee;
- **Limiting "strenuous" exercise.** Engaging in no more than five hours per week of strenuous exercise can prevent a negative effect on fertility.

For those who do not see success in their first few attempts, it can be easy to wonder if they are doing something wrong, or even if some exposure at work or work-related stress is thwarting their efforts. But at the same time, it can feel impossible to protect themselves—the way they might during a pregnancy—for an indeterminate period of time in case they have become pregnant during a given month. They may feel that changing their behavior to avoid certain work tasks may cause additional hardship on their coworkers, or that it will make others suspect that they are pregnant.

So how do veterinary professionals navigate trying to get pregnant while having exposure to workplace hazards that can reduce fertility? The preceding six chapters offer information on potential hazards and ways to mitigate those risks, and it is worth having a discussion with a healthcare provider about any specific risks and concerns to help formulate a plan. The good news is that by taking common sense precautions such as handwashing, wearing personal protective equipment when needed, avoiding unshielded radiology exposure and large doses of anesthesia, and avoiding severe exertion and stress (although many people become pregnant despite this), the risk of veterinary work contributing to infertility is low.

Another consideration when to start trying to conceive is age. Veterinarians are usually very busy with their training until at least their mid-20s and may choose to wait to start a family until after they have finished their studies and started working. Veterinary support professionals may also choose to wait until they have either finished their schooling and/or worked for a few years before trying to conceive. This may mean missing out on some of the most fertile years of life. Female fertility is at its highest from the late teens through the late 20s, and it starts to decline in the 30s, with a significant drop-off in the late 30s. Increasing maternal age also comes with an increased risk of having multiple pregnancies, chromosomal abnormalities, miscarriage, and higher-risk pregnancies.

Male fertility is at its highest between 25 and 29 years of age, and sperm quality starts to decline after this point. After 36 to 37 years of age, damage to sperm DNA can contribute to a slight increase in the risk of chromosomal abnormalities in a pregnancy. Semen volume decreases significantly around 45 years of age. As aging continues, testosterone levels decrease, leading to decreased sexual function and further decreases in sperm quality.[2] None of this means that individuals cannot naturally conceive and have healthy pregnancies in their later 30s and 40s. People over 35 years of age can and do conceive (all the time!), but it may take longer, and a higher percentage will require some sort of assistance or intervention. It is worth considering this information, where possible, when deciding on the timing of attempts to conceive.

BOX 7.1   Practice profile: Infertility and family planning in veterinary medicine

Justine Lee, DVM, DACVECC, DABT

DIRECTOR–MEDICINE/ CO-FOUNDER, VETGIRL

(Adapted from her previous writings with permission.)

Dr Justine Lee is well known in the veterinary world for her role founding and leading VETgirl, an online veterinary continuing education multimedia company. She is double boarded in emergency/critical care and toxicology, and also a mom. She has talked openly about her struggles with infertility and trying to conceive and has some important advice for would-be moms in veterinary medicine: consider not waiting!

When Justine graduated from vet school, she went straight into her internship at Angell, where she worked more than 100 hours weekly on average. From there, she did a residency at the University of Pennsylvania in Emergency and Critical Care, where she was also very busy. After that, she secured a faculty position at the University of Minnesota teaching students, interns, and residents at their veterinary school. A self-described workaholic, she threw herself into her job.

She describes getting married by 40 years of age and instantly trying to get pregnant. "Several pregnancies and losses later," she writes, "we decided to try IVF." It was then that she learned that IVF in people over 40 has a very low success rate without the use of donor eggs, which she and her husband chose not to pursue. After three years of battling infertility and multiple losses, they decided to stop and accept that they would not have any two-legged children. Six months after stopping their efforts, Justine discovered she was pregnant at the age of 43 and was able to have a healthy pregnancy and deliver a baby boy.

Justine has some advice for people in the veterinary profession based on her experiences. She knows as well as anyone that we are very goal-oriented, hard-working, and passionate about our veterinary profession and our careers. But, she counsels, "If you really want to have kids, and it's a high priority in your life, take time to prioritize it."

Her three pieces of advice?

1. Work to better manage chronic stress that can interfere with fertility;
2. Consider freezing your eggs if you see yourself putting off having children beyond your mid-30s;
3. "Stop, Drop and Roll" (i.e., *Don't* wait too long to try to get pregnant, keeping in mind there is no perfect time, and that you can still pursue your professional goals while raising kids).

As Justine says, "You can still be an awesome veterinarian, continue your career path, buy a veterinary clinic, be a specialist, and take over the world, one dog-day at a time … while being a parent."

FERTILITY TREATMENT AND ASSISTED REPRODUCTIVE TECHNOLOGY

It can be normal for some people in heterosexual relationships to take up to one year to become pregnant from the time they stop using contraceptive measures. It is usually only after trying for 1 year (or 6 months if they are ≥35 years old) that a healthcare provider would consider intervening to diagnose and treat infertility.[1]

Infertility is defined "not being able to get pregnant (conceive) after one year (or longer) of unprotected sex."[3] Female causes of infertility can include medical conditions such as polycystic ovarian syndrome, endometriosis, fibroids, and/or blocked fallopian tubes. Other causes include poor egg and/or sperm quality, which make unassisted fertilization less likely to occur.[4] It is reported that one in eight heterosexual couples will experience infertility.[5] A full discussion of the many potential causes of infertility is beyond the scope of this book, and those who are concerned are advised to consult their healthcare provider for guidance.

Apart from medical causes that prevent successful conception, there are additional reasons for fertility intervention. Some individuals choose to have a biological child without a partner. Many LGTBQI couples also wish to grow their families. In each of these scenarios, fertility treatments and/or assisted reproductive technology may help parents accomplish their goals.

Fertility treatments are a broad category of aids used to increase the likelihood of conception. Assisted reproductive technology is defined as fertility treatments in which eggs and/or embryos are handled.[6,7] Table 7.1 reviews the most commonly utilized fertility treatments along with a brief description of each, along with their known success rates and level of invasiveness.

SUPPORTING COLLEAGUES AND EMPLOYEES UNDERGOING FERTILITY TREATMENT

Fertility treatments vary in complexity and duration but can often require frequent doctor visits, bloodwork, and ultrasonograms (US). Frequent medical appointments can be very challenging to navigate with a busy work schedule. Offering understanding and flexibility to employees while they are pursuing diagnostics and treatment is an important way to help support them.[5] In addition, although fertility treatments are not always covered by insurance, some employer-provided health insurance plans have begun to cover fertility treatments, including infertility diagnosis, IVF, egg freezing, donor sperm and/or eggs, intrauterine insemination, and gestational carrier services. Offering financial coverage for these services can make an employer more competitive in attracting new employees and in retaining current employees. It can greatly improve the satisfaction and quality of life of affected employees, and it can improve equity across groups who either cannot grow a family without assistance or who would not be able to afford fertility services without employer-sponsored support. Perhaps most importantly, regardless of whether they have available benefits, employers approached by employees in need of fertility treatment should plan to "lead with empathy."[5]

HOW TO ASK FOR EMPLOYER SUPPORT WHILE UNDERGOING FERTILITY TREATMENT

There is no legal requirement to disclose attempts to conceive or fertility treatments to an employer. However, in some cases it can be helpful to disclose some basic information to a manager or HR representative to help receive any needed support.

Before requesting employer benefits and support during fertility treatment, personnel may choose to research available benefits through their employer's benefits guide or website. Some fertility benefits offered through the workplace may be accessible directly through the carrier without having to discuss it with the employer. Some employers also

Table 7.1 Summary of common fertility treatments and assisted reproductive technologies

Type of treatment	Mechanism of action	Invasiveness	Side effects	Estimated success rate
Oral medications	Increase ovulation rate	Minimal (oral medication)	Hot flashes, mood swings, pelvic pain, fatigue dizziness, risk of superovulation/ multiple births	About 19–27%; rate decreases with increasing age[4]
Injectable gonadotropins	Stimulate and regulate ovarian function	Mild (injections given at home)	Fatigue, bloating, mood swings, nausea, bruising at injection site, mild pelvic pain, or abdominal pain	See the section on IVF (usually used with IVF)[4]
IUI	Increase chances of egg and sperm interacting	Moderate (injecting sperm into the uterus)	Minimal cramping and spotting	A 3–5% increase in success rate[4]
Fertility surgery	Treat blocked fallopian tubes, fibroids, endometriosis, pelvic adhesions	High (laparoscopic surgery)	Pain, bleeding	Greater than 65% for endometriosis surgery[8]
IVF	Eggs collected and fertilized in laboratory, implanted in uterus	Very high (multiple injections, US, and surgical procedures required)	Fatigue, bloating, mood swings, nausea, bruising at injection site, pelvic pain or abdominal pain, vaginal discharge, superovulation, multiple births, ectopic pregnancy	About 12–50% of embryo transfers are successful, depending on the mother's age[4]
Donor sperm	Provides a source of good quality sperm if one is not otherwise available	Moderate (artificial insemination)	Cramping, light bleeding[9]	Varies, but frozen sperm is about 50% as successful as using fresh sperm[10]

(Continued)

Table 7.1 (Continued)

Type of treatment	Mechanism of action	Invasiveness	Side effects	Estimated success rate
IVF with donated eggs	Improves chances of conception if egg quality and/or quantity is ow[11]	Very high (see IVF)	Increased risk of pregnancy hypertension, postpartum hemorrhage, and increased risk of cesarean delivery; possible increased risk of low birth weight and preterm delivery[12]	Around 52%[13]
Embryo donation/ adoption	Embryo adopted from another family's remaining embryos from IVF	Very high (see IVF)	Increased risk of ectopic pregnancy, multiple births[14]	43–45% live birth rate[15]
Gestational carrier (surrogate)	Parents' (and/or donor) egg/sperm used for IVF, embryo implanted in gestational carrier[16]	Very high (see IVF)	Increased risk of multiple births	10–64%, depending on the age of the egg donor, and whether the eggs are frozen or fresh

IUI, intrauterine insemination; IVF, in vitro fertilization; US, ultrasonogram.

offer an employee resource group, where personnel can ask questions of coworkers (without the employer seeing it) to hear what others have done in similar situations.

When fertility treatments necessitate frequent medical visits and procedures that must be planned on short notice, it can be hard to avoid some disruptions to work schedules. For those who are in need of scheduling accommodations, they will likely need to speak with their employer or with their human resources department. It can be helpful to prepare some notes or a script ahead of time to determine what information needs to be shared and what the goals of the conversation are. This can also be helpful for those who are worried they may be overcome by their emotions when they discuss this sensitive topic.

An example script might include,

> I am currently pursuing fertility treatments. This will require me to have medical appointments every _____. I will need _____ but can make up the hours/days (or use my pto or special infertility pto if offered). I may need to step out briefly at times to receive important phone calls about scheduling and progress of the treatment. I will also need to be out of work for a few days without much warning in the next couple of weeks, and I do not have much control over the timing of this. Here are some ways I can propose to offset the time I will miss...

Those who do not wish to specify the reason can simply indicate it is for medical treatment for a non-life-threatening condition that should not affect their ability to work and give as much of an idea of the frequency and duration over which the change is needed as possible. A healthcare provider should be able to provide a letter documenting the employee's need for time off for medical procedures without revealing specific details. If employers offer specific benefits for fertility treatments, however, these may only be available when the employee discloses that they are pursuing fertility treatments.[17]

BOX 7.2   Practice profile: A veterinary neurologist's fertility journey

Erin Akin Smith, DVM, DACVIM (Neurology)

 I'd never been a woman who desperately wanted to be a mom. I always thought I would be, but it didn't happen. I eventually married at 39 years old and got pregnant naturally not long before the wedding. Six weeks before my wedding, I had a miscarriage and a D&C. I'm still not over it.

Time went on and my husband and I investigated fertility treatment. There were five emotionally exhausting failed IUIs and then we made a move from Michigan to Georgia so everything was put on the back burner. Around 2018–2019, we started to talk about having a baby again, but we were older and hesitant. I found a new reproductive endocrinologist in Georgia, but, again, life got in the way. We met with a new doctor at the same practice in 2021 and moved forward. He went over all the options—including donor egg and donor embryo. For several reasons, my husband and I chose a donor embryo and were blessed with a 3-pound, 14-ounce baby boy in May 2022. He was delivered six weeks early via urgent C-section due to preeclampsia. He then spent five weeks and four days in the NICU.

Infertility is a long and heart-wrenching journey, and it will absolutely infiltrate your work life. I am a small animal veterinary neurologist and work four days per week in a busy private practice. I chose to be private about my fertility journey from the beginning and that is a choice you will need to make—to share or not. My work schedule was flexible enough that I was able to schedule a lot (and there are a lot!) of my doctor appointments on my off days and I could block off an hour or two during a workday if I had enough notice. Be aware though, your observant team members will likely notice and figure it out. If you are more open about your journey, I'd share away and schedule your frequent doctor appointments as needed. It is up to you if you decide to share your journey with your management team. If you are comfortable, I recommend talking to colleagues/peers in your practice that have been through trying to conceive (infertility of any type) and/or pregnancy.

My typical workday didn't change much—I still charted on my computer in my office and saw patients in exam rooms. I continued to work four days per week. If that is difficult for you, consider backing down to three days per week or something else that works for you. When possible, I recommend having your technicians and assistants do all the lifting of your patients onto the exam table. This worked well for me; I was also able to get on and off the floor relatively easily for a long time, so I performed many exams on the floor. Wash your hands frequently and wear gloves with every patient. I also requested that I not be scheduled any pets over 100 pounds. I was still able to perform anesthetized procedures, but I would make sure everything was set up prior to me entering the room to avoid being exposed to any leaks. Your staff will be invaluable in this regard. Remember, ask for help. You are not alone.

ADOPTION

Adoption is pursued for a variety of different reasons, some of which have nothing to do with infertility, to help families grow. While it is chosen by some parents whose efforts to conceive naturally or through fertility treatments have been unsuccessful, single parents by choice may consider adoption, as may some LGTBQ+ couples. Some parents may be drawn to the idea of adoption to help give a child in need a home. Others may be in the position to adopt one or more family members in need of care. Some stepparents will adopt their stepchild to formalize their relationship. Adoption can take many forms; they can be open or closed, domestic or international, infant or older child.[18]

Although in the past many adoptions were closed, meaning that the birth parents signed away their rights and had no further contact with or news about their biological child, the trend in recent years is more toward open adoptions. Open adoptions allow for the birth parent(s) and the adoptive parent(s) to meet, stay in contact, and allow the child to grow up knowing their biological parent(s). This can help adopted children have a sense of where they came from, to know their own adoption story, and to have better access to health information and family history.[19]

CHALLENGES OF ADOPTION

As parents who have been through it will attest to, adoption is not without its possible challenges. One of the most common is the cost involved, which may include legal fees, agency fees, in some cases living expenses and medical expenses for the birth parent during pregnancy, fees for background checks and home inspections, and travel costs; and for international adoptions, passport fees and fees imposed by foreign governments. In some cases, the costs can exceed 30,000 USD.

Aside from the cost, couples who prefer an infant of their own ethnicity may find that they have extended wait times and competition from other hopeful parents. Those who adopt internationally may find that they also have prolonged wait times or that they even have to stay in the host country for an extended period of time before they are allowed to travel with their adopted child.[20]

Parents who adopt a child who has already experienced loss, abuse, abandonment, or other traumatic circumstances may find they have additional challenges in terms of establishing trust with their child.[21]

SUPPORTING THE ADOPTIVE PARENT

Whether they have been open about it or not, parents who are pursing adoption may have already faced a long and arduous process. There can be many challenges associated with the adoption process, sometimes including extensive wait times and diminishing financial resources. Parents who are pursing adoption may need additional time off work for meetings with agencies or for travel to meet prospective birth families. In some cases, they may need to travel on short notice if a child becomes available for adoption. Some parents may also choose to foster before adopting, and they may need accommodations to their schedule to be able to care for any foster children.

The consideration given the adoptive parent should match the consideration given the parent pursing fertility treatments, the parent who has just given birth, or the parent who has just experienced a loss. Adoption assistance benefits in the workplace are becoming more commonplace. Some benefits offered by employers can include reimbursement for eligible expenses, paid leave, scheduling flexibility, employee assistance programs, and access to educational resources about the adoption process. These benefits can greatly support the quality of life of employees who take advantage of them and can be used as a powerful recruiting and retention tool in a competitive job market.[22]

PREGNANCY LOSS

About 25% of all known pregnancies end in some form of loss.[23] These include miscarriage, stillbirth, and other causes. Grieving parents can be supported in many ways, both personally and professionally.

MISCARRIAGE

Miscarriage (also called spontaneous abortion) is defined as the loss of a pregnancy before 20 completed weeks' gestation.[24,25] It is estimated that about 10% of known pregnancies (30% in people >40 years of age) end in a miscarriage. About half of those are the result of chromosomal defects that prevent the embryo from properly forming and/or growing. Miscarriage is almost always random, unpredictable, and not the fault of either parent. Work, stress, exertion, or even most injuries are unlikely to result in a miscarriage.

According to the Mayo Clinic, signs and symptoms of a miscarriage may include spotting or bleeding, cramping and/or pain in the lower abdomen or back, and sometimes the passage of fluid and/or tissue. It is important to note that not all bleeding or cramping is indicative of a miscarriage. Some people will experience this and go on to have a healthy pregnancy.[26] Other less common signs include fever, chills, tenderness in the lower abdomen, and foul-smelling vaginal discharge. These signs are associated with a septic miscarriage, which can be life threatening to the mother.[26]

Yet other people will experience a blighted ovum (in which an embryo never forms inside the yolk sac) or intrauterine fetal demise (or missed abortion, in which the embryo forms and then dies with no outward signs of a problem). It is very likely that no signs or symptoms would be noted in either of these types of miscarriage.

Depending on the circumstances, a miscarriage may be suspected based on signs and symptoms. Bloodwork is sometimes performed to monitor human chorionic gonadotropin levels to see if they are increasing as they would be expected to do in a healthy pregnancy. Often, an US is used to confirm the miscarriage (lack of an embryo or lack of a heartbeat). Sometimes, particularly in the very early weeks of pregnancy, an US must be repeated after waiting a period of days to be certain that the embryo is not growing as expected.

In some cases, no treatment is needed, as the body sheds the tissue completely. This is particularly true in very early miscarriages and with a blighted ovum. In other cases, it may be necessary to induce the passage of the tissue and/or physically remove it from the mother's body to avoid severe illness in the mother. Oral medications are sometimes prescribed to stimulate uterine contractions. This may need to be followed by a suction dilation and curettage (D&C), a procedure in which the cervix is dilated, and the contents of the uterus removed. This is usually done in a hospital setting. In some cases, a mother and their provider may decide to pursue the D&C without attempting other treatments first.

STILLBIRTH

Stillbirth is defined as the death of a fetus after 20 weeks of pregnancy. The causes in most cases are unknown, but stillbirth can be the result of genetic abnormalities, problems with the health of the placenta, infections or other medical conditions affecting the mother, complications during labor and delivery, and problems that affect the umbilical cord.[27] Stillbirth occurs in about 1 in 175 births in the United States.[28] A pregnant individual may notice the absence of fetal movement, if they have reached the stage in pregnancy where they were able to appreciate regular movement.[27] Spotting or bleeding may also be noted.[29] Stillbirth may be suspected after the fetal heartbeat is not heard on a Doppler monitor, and that diagnosis is often confirmed by US. The cause of the stillbirth may be difficult to determine.[27] Depending on the stage of pregnancy and the health of the mother, dilatation and evacuation (D&E), labor induction, or other treatments may be offered.[27]

BOX 7.3   Practice profile: Pregnancy loss

Alycia Cullen, DVM

I joined my second job after veterinary school right after my maternity leave for my first son. It was a full-time emergency medicine at a very busy 24-hour hospital. Staff were often scarce and untrained, leaving me to do a lot of physical and non-medical tasks throughout the shift.

After over two and a half years of working full-time night positions, my husband and I decided to try for another pregnancy. I luckily became pregnant immediately and dutifully told my medical director in order to explain my reluctance to take X-rays or handle aggressive pets. I was told that "X-rays were not that harmful if [I] wore the gowns properly, and we won't be able to bring in more staff just to help" me.

When I discovered my missed miscarriage at 12 weeks, my husband and I were devastated. We were given options of passing the embryo at home naturally or scheduling a D&C surgical removal. We elected a surgical removal, and the first time available was offered the morning after a scheduled overnight shift. My doctor recommended that I rest until then and warned me of potential bleeding and natural passage of the tissues before the D&C. When I called my medical director to notify her of the loss and plan, she was very sympathetic and kind, but she stressed that I was responsible for covering the shifts unless someone, who I needed to find myself, would cover it.

I felt exposed and betrayed. It seemed I was expected to explain my horrible situation to other colleagues in order to beg for time off to grieve and safely pass my pregnancy. When I pushed back about not wanting to come in in case I passed the tissues on shift, she told me: "I have not had a miscarriage myself, but I have some friends who have, and they said it was basically a heavy period. You could wear a pad just in case." Then, while my shifts were 6:00 pm to 8:00 am, I was expected to be at the hospital for pre-op at 5:00 am. I told her that I would not be able to complete that shift to prepare myself for surgery, and I was told "I'll try to come in around 4:15 am to relieve you, so that you can go straight to the hospital. It might be great anyway, because you will be so tired, the anesthesia will work well."

It was through that experience that I knew my job and my leadership did not care for me, my health, or my mental health over the coverage of a hospital shift. I ended up having two additional early miscarriages before I chose to leave the practice and move on to another passion. I now have two beautiful rainbow babies and live in disbelief that I was willing to work for someone with those priorities and work ethic.

OTHER TYPES OF PREGNANCY LOSS

MOLAR PREGNANCY

Chromosomal defects can also result in a molar pregnancy, which is a nonviable pregnancy in which the placenta does not grow normally and can cause a secondary cancer in the mother.[30] Molar pregnancies are uncommon, but they can be very dangerous for the mother. Signs and symptoms include dark brown to bright red vaginal bleeding in the first trimester, pelvic pressure or pain, severe nausea and vomiting, and, sometimes, passage of grapelike cysts from the vagina. A molar pregnancy must be removed. Cancer may still develop in some people after the removal, and they may need chemotherapy.[30]

ECTOPIC PREGNANCY

Ectopic pregnancy is a pregnancy in which the embryo has implanted somewhere other than in the uterus, usually in a fallopian tube. This pregnancy will not be viable, and it represents a danger to the health of the mother because of severe bleeding. There may be no early signs or symptoms of an ectopic pregnancy. Over time, light vaginal bleeding and pelvic pain can become noticeable. If the fallopian tube begins to leak blood, symptoms such as pain in their shoulder(s) or an urge to defecate may be appreciated. In severe cases, the fallopian tube can rupture, causing heavy bleeding into the abdomen. Lightheadedness, fainting, and shock (in advanced cases) may be noticed.[31]

There are some risk factors that can make an ectopic pregnancy more likely. These include any inflammation or infection of the uterus and fallopian tubes, previous history of surgery of the fallopian tubes, previous ectopic pregnancy, smoking, and certain forms of birth control such as intrauterine devices.[31]

PREMATURE BIRTH

While advances in neonatal medicine and care have improved outcomes, babies who are born significantly before they have reached full term can be at increased risk for infant death. In 2020, 16% of infant deaths were associated with preterm birth or low birth weight. Possible causes of premature birth can include a pregnancy with multiple babies, intervals of less than 6 months between pregnancies, infections of the amniotic fluid or uterus, some injuries, and problems with the cervix, uterus, or placenta. Other causes can include smoking, abusing drugs or alcohol, and severe stress. The risk can also be higher in parents over age 35 and in Black and Native American mothers for reasons that are not fully understood.[32]

RECOVERY AND GRIEVING

The physical recovery from a pregnancy loss can vary from short (hours to days in an early miscarriage) to much longer (weeks to months with a late-term loss or if the mother had other health challenges). The emotional recovery can vary just as much. All reactions, from relief to indifference to profound grief, are normal emotional responses.

For those who experience it, grief can include feelings of guilt, anger, disbelief, or despair. Seeing a mental health counselor can be very helpful in working through all of the emotions associated with the loss. Some communities and hospitals offer support groups where parents can interact with others who have experienced a similar loss. Many families will find comfort in finding ways to honor and remember their unborn child through a memorial ceremony or funeral, sometimes including cremation and/or burial. Grieving parents can find other ways to commemorate remembrance such as jewelry, artwork, keychains, teddy bears, blankets, and more. Some may find solace in *creating* such items.

TRYING AGAIN

Depending on the cause of the loss, how far along the pregnancy was, and any other complicating factors, a healthcare provider may recommend waiting a period of time before trying to conceive again. Many parents will find they need this time to recover somewhat emotionally before looking forward to a future pregnancy. Pregnancy after a loss can be associated with more anxiety for some people, as they worry that they might experience the same type of loss again. Depending on the circumstances, healthcare providers may recommend additional monitoring during subsequent pregnancies. This is not always necessary, however.

SUPPORTING PARENTS AFTER PREGNANCY LOSS

Just as everyone's experience of grief is different, so will their needs be as they begin to process their loss. Some parents will need privacy and time alone before they are ready to receive even the most well-intentioned care and service from friends, family, and coworkers. Others will be open to receiving more support right away. It can be helpful to ask the parents what they need.

Some ways to support the grieving parents include offering to care for other children in the home, dropping off food, or asking the parents if they need help putting away any baby items in the home. For parents who had already named their baby, it can be very comforting to hear others refer to their baby by name.

Another very important way to support grieving parents (both the birth parent and non-birthing parents) is by offering paid time off after a pregnancy loss. This type of paid time off has not typically been commonplace in the United States, but there are some countries where it is now required. These include India, the Philippines, South Korea, and New Zealand. Some individual companies in the United States have followed suit, offering anywhere from 15 days to 4 weeks of paid leave after pregnancy loss, whether it was the employee who suffered the loss or their spouse or gestational carrier. Some of these companies also offer paid days off after a failed fertility treatment, recognizing that there can be significant grieving associated with this loss as well.[33]

PREGNANCY TERMINATION

A pregnant person may decide with their healthcare provider to terminate a pregnancy for a variety of reasons. While some of them have to do with a detected abnormality in the pregnancy or a maternal health condition, many others do not. In some cases (but certainly not all), the termination of a pregnancy can represent a loss to the parent(s) every bit as much as a miscarriage or stillbirth. For others, it provides necessary reproductive control. Either way, deciding to terminate and pursing the termination can be emotionally and/or logistically challenging.

Based on the state and/or country of residence of the mother, terminating a pregnancy may be easily accomplished, very tightly regulated, or even illegal. As such, some parents will need to travel to safely and legally terminate a pregnancy. This can create financial hardship and lost time at work, in addition to the stress or grief sometimes associated with the decision to terminate.

SUPPORTING THE PARENT PURSUING ABORTION

In the United States, charitable organizations exist to help pay for travel and other related expenses for the purpose of pregnancy termination. Some also provide free abortions or funds to cover the cost of an abortion if health insurance does not cover it. Some employers also offer financial assistance for travel, medical care, and/or paid time off for pursuing an abortion for employees and, in some cases, their family members.[33]

Even when an employer advertises that they offer assistance or support related to an abortion, it can be daunting to consider discussing something so personal at work. Unfortunately, in some cases, the only way for an individual to access abortion-specific benefits will be to advise their employer of their needs. Before doing this, it is advisable to review the laws surrounding abortion in the state, province, and/or country in which a pregnant person works and resides. If abortion is not currently legal there, it will be important to determine if laws also make it illegal to assist a person with their abortion in any way. Then it will be easier to decide if disclosing a planned abortion to an employer feels safe.

When an employee has concerns about sharing their need for an abortion with their employer and the possible ramifications, experts recommend looking for any non-abortion specific benefits first. These can include regular paid sick days, which typically do not require any kind of medical justification. Some employers may offer a general reimbursement for any reproductive care not covered by health insurance, which may include travel to obtain that care. When documentation from a healthcare provider is required, this does not necessarily have to include the type of care being provided.

In situations where advising an employer of an abortion feels necessary or desired, some experts recommend not volunteering private health information to an employer, but instead providing only general information unless the employer asks for more details. This is because the Americans with Disabilities Act (ADA) in the United States requires that employers in workplaces of more than 15 employees keep any health information private if the information was provided at the request of the employer. This same protection does not necessarily apply to information that is volunteered by the employee. Examples of this include advising an employer of a need for "medical treatment" for a "health issue" or "medical emergency," and then providing more details if desired when or if an employer asks for them.

It can also be helpful to provide the following statement to an employer in writing:

> Please keep this private, protected medical information confidential, and do not share it with my coworkers, supervisor, or anyone else who does not have a

legitimate reason to know it. If you need to share it beyond Human Resources for any reason, please contact me first. Please keep this information separate from my general employee records, as may be required by state and federal law. Thank you for protecting my privacy.[34]

TAKE-HOME POINTS

- Success rates of unassisted conception can be maximized by monitoring ovulation, maintaining a healthy body weight, and having intercourse daily to every other day until a woman's ovulation window has closed;
- Other helpful tips include avoiding alcohol and smoking, reducing caffeine, and limiting strenuous exercise to five hours a week or less;
- Infertility can have multiple causes and is usually only diagnosed after at least one year of trying to conceive unsuccessfully;
- A variety of fertility treatments exist to help make conception more likely and to help single parents and LGBTQ+ parents grow their families;
- Although it is not for everyone, some families will benefit from various forms of adoption;
- Pregnancy loss affects up to one in four known pregnancies. Most pregnancy losses are caused by chromosomal abnormalities and cannot be prevented;
- Providing flexible scheduling, paid leave, and/or other financial support can help team members who are navigating fertility treatments, adoption efforts, pregnancy loss, or pregnancy termination to feel supported and to get the care they need. These types of employment benefits are standard in some countries and are being offered by more and more employers the world over.

REFERENCES

1. Mayo Clinic. Getting pregnant. December 11, 2021. Accessed October 1, 2022. https://www.mayoclinic.org/healthy-lifestyle/getting-pregnant/in-depth/how-to-get-pregnant/art-20047611
2. Path Fertility. Age affects male and female fertility. Accessed October 12, 2022. https://pathfertility.com/age-affects-male-and-female-fertility/
3. Centers for Disease Control and Prevention. Infertility FAQs. March 1, 2022. Accessed October 1, 2022. https://www.cdc.gov/reproductivehealth/infertility/index.htm
4. Conrad, M. A beginner's guide to fertility treatments. June 15, 2022. Accessed September 30, 2022. https://www.forbes.com/health/family/fertility-treatments-guide/
5. Ravishankar RA. Does your employer offer fertility and family planning benefits? March 10, 2022. Accessed October 12, 2022. https://hbr.org/2022/03/does-your-employer-offer-fertility-and-family-planning-benefits
6. Centers for Disease Control and Prevention. What is assisted reproductive technology? October 8, 2019. Accessed September 30, 2022. https://www.cdc.gov/art/whatis.html

7. American Society for Reproductive Medicine. Assisted reproductive technologies. Accessed September 30, 2022. https://www.asrm.org/topics/topics-index/assisted-reproductive-technologies/

8. Kofinas G. What to expect for your fertility after endometriosis surgery (laparoscopy). December 1, 2021. Accessed October 5, 2022. https://www.kofinasfertility.com/patient-info/fertility-after-endometriosis-surgery

9. WebMD. Infertility and artificial insemination. August 1, 2021. Accessed October 5, 2022. https://www.webmd.com/infertility-and-reproduction/guide/artificial-insemination

10. Bradley S. I'm thinking of using a sperm donor...How does it work? Accessed October 5, 2022. https://cofertility.com/how-does-a-sperm-donor-work/

11. Weigel G, Ranji U, Long M, Salganicoff A. Coverage and use of fertility services in the U.S. September 15, 2020. Accessed October 5, 2022. https://www.kff.org/womens-health-policy/issue-brief/coverage-and-use-of-fertility-services-in-the-u-s/

12. Shah A, Parisaei M, Garner J. Obstetric complications of donor egg conception pregnancies. *J Obstet Gynaecol India*. 2019;69(5):395–398.

13. Pacific Fertility Center Los Angeles. IVF with donor eggs: What to expect, success rates, and more. September 10, 2019. Accessed October 5, 2022. https://www.pfcla.com/blog/donor-egg-ivf

14. Tennessee Reproductive Medicine. Embryo donation. Accessed October 6, 2022. https://trmbaby.com/library/donation-surrogacy/embryo-donation/

15. Progyny. Embryo donation benefits and associated costs. Accessed October 5, 2022. https://progyny.com/education/ivf-facts/embryo-donation-costs/

16. Fuller E. Gestational carriers (surrogacy). Accessed October 12, 2022. https://www.babycenter.com/surrogacy#articlesection12

17. Mate Fertility. How to tell your boss you're going through fertility treatment. July 20, 2021. Accessed March 21, 2023. https://matefertility.com/how-to-tell-your-boss-youre-going-through-fertility-treatment/

18. University of Washington. Adoption. Accessed October 12, 2022. https://depts.washington.edu/triolive/quest/2007/TTQ07039/textsite/typesofadoption-t.htm

19. American Adoptions. The pros and cons of adoption. What to know before choosing domestic adoption. Accessed October 12, 2022. https://www.americanadoptions.com/adoption/advantages_of_adoption

20. American Adoptions. Comparing the costs of domestic, international and foster care adoption. Accessed October 12, 2022. https://www.americanadoptions.com/adopt/the_costs_of_adopting

21. Adopt US Kids. About adoption from foster care. Accessed October 12, 2022. https://www.adoptuskids.org/adoption-and-foster-care/overview/adoption-from-foster-care

22. Soronen R. Adoption friendly benefits in the workplace: It is the right thing to do. July 1, 2018. Accessed October 12, 2022. https://adoptioncouncil.org/publications/adoption-advocate-no-121/

23. Greves C. Pregnancy loss: 1 in 4. October 30, 2018. Accessed October 12, 2022. https://www.winniepalmerhospital.com/content-hub/pregnancy-loss-1-in-4

24. American College of Obstetricians and Gynecologists. Early pregnancy loss. FAQs. Accessed September 30, 2022. https://www.acog.org/womens-health/faqs/early -pregnancy-loss

25. MedlinePlus. Miscarriage. Accessed September 30, 2022. https://medlineplus.gov/ ency/article/001488.htm

26. Mayo Clinic. Miscarriage. Accessed October 2, 2022. https://www.mayoclinic.org/ diseases-conditions/pregnancy-loss-miscarriage/symptoms-causes/syc-20354298

27. American College of Obstetricians and Gynecologists. Stillbirth. Accessed September 30, 2022. https://www.acog.org/womens-health/faqs/stillbirth

28. Centers for Disease Control and Prevention. What is Stillbirth? September 29, 2022. Accessed October 5, 2022. https://www.cdc.gov/ncbddd/stillbirth/facts.html

29. Stanford Medicine Children's Health. Stillbirth. Accessed October 5, 2022. https:// www.stanfordchildrens.org/en/topic/default?id=stillbirth-90-P02501

30. Mayo Clinic. Molar pregnancy. Accessed October 5, 2022. https://www.mayoclinic .org/diseases-conditions/molar-pregnancy/symptoms-causes/syc-20375175

31. Mayo Clinic. Ectopic pregnancy. Accessed October 5, 2022. https://www.mayo-clinic.org/diseases-conditions/ectopic-pregnancy/symptoms-causes/syc-20372088

32. Mayo Clinic. Premature birth. Accessed March 24, 2023. https://www.mayoclinic .org/diseases-conditions/premature-birth/symptoms-causes/syc-20376730

33. Johnston K. A wave of organizations are offering paid leave for pregnancy loss. February 4, 2022. Accessed October 5, 2022. https://www.bostonglobe.com/2022 /02/04/business/boston-law-firm-joins-wave-organizations-offering-paid-leave -pregnancy-loss/

34. Work Life Law. Abortion, pregnancy loss & your rights at work: Frequently asked questions. Accessed March 21, 2023. https://pregnantatwork.org/faq-abortion -pregnancy-loss-and-work/#1658946089958-7ef592e5-4907

8

"Was this planned?"

Announcing pregnancy at work

When I became pregnant with my first child just three months out of vet school, I thought I would be fired from my first job. If I wasn't fired, I thought, my new employer would surely think badly of me for being so "unprofessional" to get pregnant right after starting my first job. I felt the need to promise them that I would return after having a baby, and that I would still be a good veterinarian. Thankfully, I was met with reassurance that my job would be protected and understanding that my needs might change. I still felt that my pregnancy was something I should keep to myself and not announce to anyone who did not "need to know." Throughout subsequent pregnancies, I experimented with how and when to deliver my news to my employer and coworkers, not with any sense of fear of repercussions, but rather with a desire to keep my pregnancy private for longer. I thought that if I had a miscarriage, I wouldn't want anyone to know. Surprisingly, I found the most freedom and sense of ease when, during my last pregnancy, I decided to announce it to both management and staff very early in my first trimester. This freed me from worrying about others finding out or gossiping about my pregnancy, and it helped me to better advocate for the protections I needed.

DOI: 10.1201/9781003406907-8

There are many factors that can affect when and how a pregnant veterinary professional decides to announce their pregnancy at work. For employees, there can be concerns about how they will be perceived and/or treated once their pregnancy has been declared. For some, they may be looking for a job or may have just started at a job when they discover they are pregnant, and they worry that their pregnancy will endanger their prospects in their new role. For pregnant individuals in all roles, there can be a desire to withhold their pregnancy news until the risk of pregnancy loss has decreased. However, there can be significant benefits to disclosing a pregnancy early on in terms of access to workplace accommodations, more informed planning of maternity leave, and the moral support of management and coworkers throughout pregnancy. Not all countries are equal in terms of the workplace protections offered, but the majority of countries do have laws in place to protect pregnant individuals from discrimination and harassment. The International Labour Office of the United Nations and other organizations continue to advocate for better laws and better enforcement of current laws. In this chapter, the pregnancy protection laws of several countries are reviewed. A discussion of the timing of the announcement of pregnancy, ways to announce or disclose pregnancy at work, and tips for employers and coworkers showing appropriate responses to a pregnancy disclosure are included.

LAWS PROTECTING PREGNANCY AND RELATED CONDITIONS FROM DISCRIMINATION

THE UNITED STATES

According to the US Equal Employment Opportunity Commission (EEOC),

> If you are pregnant, have been pregnant, or may become pregnant, and if your employer has 15 or more employees, you are protected against pregnancy-based discrimination and harassment at work under federal law. You may also have a legal right to work adjustments that will allow you to do your job without jeopardizing your health.[1]

In the United States, these rights and protections are spelled out in the Pregnancy Discrimination Act (PDA), which is an amendment to Title VII of the Civil Rights Act of 1964, the American Disabilities Act (ADA), and the Pregnant Workers Fairness Act (PWFA). In addition to protecting those who are pregnant, who could become pregnant, or who were pregnant, the PDA and ADA also protect those who have a medical condition related to pregnancy, including breastfeeding/lactation, those who choose to pursue treatment for infertility, those who choose to use contraception, and those who had an abortion, who are considering having an abortion, or who have chosen not to have an abortion.[1,2]

Discrimination can include being fired, being rejected from a job or a promotion, being given "lesser assignments," or being forced to take leave. In other words, an employer cannot force an employee to take leave because they believe the work would pose a risk to a pregnant employee. Likewise, an employer is legally bound to stop any harassment directed toward pregnant employees and prevent it from continuing to happen.[1] These federal laws also prevent employers from asking an employee if they are pregnant or plan to become pregnant.[3]

The PDA dictates that if an employer gives accommodations to other nonpregnant employees who have special needs or limitations, they must also provide them to pregnant employees who request them. The PWFA, newly passed at the time of this writing, requires employers to provide "reasonable accommodations" to a pregnant employee's "known limitations related to pregnancy, childbirth, or related medical conditions."[4] These can include changes in schedule and/or breaks, more opportunities to sit or stand, ergonomic office furniture, avoiding certain shifts (like night shifts), "elimination of marginal job functions" (i.e., tasks that can be easily done by others and are not essential to that individual's role), and even permission to work from home (where practical).[1]

Some examples of possible accommodations in veterinary clinical practice include avoiding exposure to radiation (including dental radiographs); not recovering patients who have had inhalant anesthesia; not restraining very large, active, or fractious patients; avoiding certain higher-risk procedures such as joint injections in horses; avoiding handling fecal samples; not handling certain drugs or chemicals; having the assistance of a technician, especially for large animal veterinarians who might not normally have one; having the ability to sit during at least a portion of the workday; and possibly reducing or changing hours. Pregnant employees may also wish to avoid seeing patients who may be infected with certain zoonotic diseases. During the COVID-19 pandemic, some pregnant veterinary professionals chose to have less face-to-face client interaction to limit their potential exposure, and instead cared for their patients through curbside-only appointments.

While the PDA and the PWFA cover pregnancy and childbirth broadly, the ADA covers pregnancy-related health conditions such as preeclampsia, sciatica, anemia, cervical insufficiency, gestational diabetes, depression, and others that meet the definition of "disability." To be considered a disability, a condition does not have to be so severe that a person cannot work at all. Rather, it must "substantially limit" at least one "major life activity" (eating, sleeping, bending, walking, reaching, standing, concentrating, etc.) or "major bodily function" (cardiovascular, digestive, genitourinary, bowel, bladder, neurological, etc.). A disability causes activities to be "more difficult, uncomfortable, or time-consuming to perform compared to the way that most people perform them," even if they are not impossible to perform. When a disability is recognized and reported to an employer, accommodations that don't present a significant financial hardship or difficulty to the employer should be offered. In cases where an employee cannot continue to work due to their pregnancy and/or pregnancy-related disability, unpaid leave is also considered to be an accommodation.[1]

The Family and Medical Leave Act (FMLA) can provide additional unpaid job protection to eligible employees, including up to 12 weeks per 12-month period of unpaid, job-protected leave that includes the continuation of group health insurance for "a serious health condition that makes the employee unable to perform the essential functions of his or her job." Using all or some of FMLA leave during pregnancy, however, will reduce the remaining time covered by the FMLA available for healing after delivery and bonding with the baby during that same 12-month period.[5]

Pregnant employees who feel they need an accommodation should discuss this with their supervisor or manager as soon as possible. Employees cannot be fired or otherwise penalized for requesting an accommodation. If an employee is requesting an accommodation for a pregnancy-related medical condition, an employer may request a letter documenting this from the employee's health care provider.

Unfortunately, it is true that employers with fewer than 15 employees may not offer as many protections for pregnant employees under federal law. There may, however, be other state or local protections or policies in place at the employer level. The EEOC recommends that pregnant employees who are concerned about their work protections contact their local Women's Bureau office to get guidance and assistance.[1] This is a branch of the federal government that "champions policies and standards that safeguard the interests of working women, advocates for the equality and economic security of women and their families and promotes quality work environments." The Women's Bureau maintains information about equal pay and pay transparency protections, access to paid leave by state, childcare prices, racial disparities in wage inequality, and a listing of workplace protections for pregnancy and breastfeeding at the federal level and by state. They also provide resources for those who may have experienced discrimination so that they can find support.[6]

CANADA

According to the Canadian Human Rights Commission, the Canadian Human Rights Act prohibits "discrimination related to pregnancy." This definition of discrimination includes "negative treatment, refusal to hire or promote, termination of employment, or harassment." As with US laws, this protection is extended to fertility treatments, family planning, pregnancy-related medical conditions, pregnancy loss and conditions related to pregnancy loss, abortion and conditions related to abortion, pregnancy, childbirth and recovery, maternity leave, and breastfeeding. Also included are pregnancy as a surrogate and placing a newborn baby up for adoption. An employer is required to provide needed accommodations to a pregnant employee, as long as they do not impose "undue hardship" on an employer. The employer should follow restrictions ordered by a doctor and be mindful of the accommodation requests of the employee, but they should not assume that a pregnant employee cannot perform a task because of their pregnancy.[7]

UNITED KINGDOM

According to the government of the United Kingdom, "pregnant employees have 4 main legal rights: paid time off for antenatal care, maternity leave, maternity pay or maternity allowance, and protection against unfair treatment, discrimination or dismissal." Antenatal care is defined as both medical appointments and any classes, such as parenting classes, that have been recommended by a healthcare professional. A non-birthing partner is also entitled to unpaid time off to attend two antenatal appointments. Employees are required to tell their employer about their pregnancy at least 15 weeks before the start of the week in which their baby is due, or as soon as possible if they did not know they were pregnant at that point. Employers must conduct a risk assessment and work with the employee to take "reasonable steps" to remove risks or suspend the employee with full pay if they cannot do this.[8]

The Advisory, Conciliation and Arbitration Service states that discrimination—including dismissal, refusing to offer a job, changing pay or other terms of a contract, requiring employees to work while on maternity leave, or preventing them from returning to work due to breastfeeding—is prohibited. This protection prevents discrimination because of pregnancy, any medical condition related to pregnancy, maternity pay, and maternity leave. It extends from the beginning of pregnancy until maternity leave ends, the employee returns to work, and/or they leave their job.[9]

AUSTRALIA

The Australian government states that laws protect employees from being "fired, demoted or treated differently from other employees because they are pregnant." Pregnant employees are entitled to sick leave if they have a pregnancy-related medical condition, and they may also be eligible for unpaid special maternity leave for pregnancy-related illness, miscarriage, or termination. Compassionate leave may also be available to both the pregnant employee and their partner for miscarriage or stillbirth. All pregnant employees, even those who are not eligible for unpaid parental leave, are entitled to be transferred to a "safe job" if their regular role is not safe to do during their pregnancy. Employees who are moved to a safe job will receive the same pay rate, hours, and benefits that they had in their previous role, unless a new arrangement is agreed upon. Employers may ask for a medical certificate documenting the need for an accommodation. Employees who cannot be given a safe job and who are eligible for unpaid parental leave are also eligible for paid "no safe job leave." For those not entitled to unpaid parental leave, they can still take unpaid "no safe job leave."[10]

OTHER COUNTRIES

In a report from the International Labour Organization, an agency of the United Nations charged with "advancing social justice and promoting decent work," 20 of 165 of the countries that reported information had no legal protections for pregnant individuals in the workplace.[11] In the countries where legal protections are present, they vary in terms of comprehensiveness and clarity. There is also a lot of variation in terms of the length of time over which protection to the employee is extended. The Republic of Moldova, for example, provides legal protections against wrongful dismissal from pregnancy extending until the child is six years old. Many other countries, including Chile, Panama, Bolivia, Angola, Somalia, and Vietnam, provide protection from discrimination based on pregnancy from the start of the pregnancy until one year after the child is born (or one year after the end of maternity leave in Chile and Panama). Some countries continue the protection until the end of the "nursing period," which may be a different length of time in each country.[12]

There are more limited protections offered in other countries. For example, in Botswana, Cambodia, Fiji, Lebanon, Lesotho, Libya, Malaysia, Niger, and Paraguay, legal protection from dismissal because of pregnancy only extends from the beginning of pregnancy until the end of maternity leave. In Egypt, only the maternity leave itself is protected; in Chad, only the pregnancy is protected.[11]

Countries also vary in terms of their laws regarding "dangerous or unhealthy work" for pregnant or lactating people. Seventy-eight percent of the countries surveyed had some specific prohibition against work that could be hazardous to the health of a pregnant or lactating employee or their child. This can include prohibiting overtime work, "arduous work," or anything "injurious to pregnancy, childbirth, nursing and related matters." Countries including Bosnia and Herzegovina, Brazil, Cambodia, Cape Verde, Denmark, Guatemala, Kenya, Myanmar, and Romania have no legal protections against hazardous work during pregnancy.

At least 25 countries have policies requiring a workplace to perform a "risk assessment" as soon as they are informed that an employee is pregnant, and in some countries, employers are required to provide the pregnant employee information on "work-related risks during pregnancy and breastfeeding periods" within a short period of time after an employee reports their pregnancy. The International Labour Organization recommends having laws in place that prevent pregnant employees from being forced to do dangerous work, but not prohibiting them from performing any task because of their pregnancy. This allows the individual the protection to avoid hazardous tasks if it is in their best interest, while allowing them to continue to perform their work if they feel comfortable doing so and return to those tasks as soon as they are ready.[11]

WHEN DISCRIMINATION HAPPENS

Even with protections in place, pregnancy discrimination does still exist. An estimated 30,000 pregnant workers were losing their jobs every year in 2014 in the United Kingdom because they were pregnant, and in 2018, this number was quoted to be 54,000.[13,14] In Korea, one-third of pregnant employees take little to no maternity leave for fear of discrimination. Pregnancy discrimination complaints are also high in Australia, Costa Rica, and the Dominican Republic. Even in the United States, pregnancy discrimination claims increased by 31% between 2005 and 2010.[13]

Many countries, however, do have provisions in their laws that provide for compensation to affected employees when a violation has been proven.[11] Employees who feel they have experienced discrimination related to their pregnancy should strongly consider reporting it. This may not only provide reconciliation for their own case, but also improve conditions for future employees. Employees in the United States who feel that their rights under the PDA, PWFA, or ADA have been violated should file a charge with the EEOC within 180 days of the violation, and employers cannot retaliate against employees for doing so.[1] In Canada, employees can file a complaint with the Canadian Human Rights Commission. In the United Kingdom, employees can file a claim with the Advisory, Conciliation and Arbitration Service within 90 days of the discrimination.[15] In Australia, claims are made to the Australian Human Rights Commission.[16]

BOX 8.1 Practice profile: The power of the employer response

(Name withheld), DVM

When I think about the relationship between employer and employee, announcing a pregnancy and the support you receive during your pregnancy is just so critical to retaining employees and building a strong relationship. I remember feeling so nervous to tell my employer that I was pregnant; I was near tears. I knew it would make their lives so much more difficult and there were a lot of unknowns about how I'd be able to perform at work, when I'd need to stop work, what this meant for my clients, etc. My two pregnancies were also really different, both in employer reaction and health. My first pregnancy was healthy with almost no ill effects of pregnancy. Employer support was really where things lacked. I shared the news with my employer as an associate and they immediately said, "Well, that sucks. It sucks for us big time. Thanks for ruining my day." The practice owners then failed to make medical accommodations for me and left me to figure it out on my own while they went on vacation. My employer disclosed my pregnancy to staff and clients before I was ready and made terrible comments about how people who wanted to get pregnant needed to earn that right. They also made no effort to make themselves aware of the legal requirements around pregnancy, and I was left to let office staff know what I was and wasn't doing. I was their top associate, and they seemed surprised when I left.

My second pregnancy was such a stark contrast. I was doubly nervous to tell my boss this time around after the last experience. My new boss heartily congratulated me and gave me a hug. We made a plan for the work I did, and an emergency plan to ensure work was handed over in the event of medical issues, and a plan to disclose the news with my wishes in mind. Throughout, I felt supported and like pregnancy and having children was something that was permitted, accepted, and a happy event, even if it meant challenges for my employer. I was quite sick during this pregnancy and had to stop work early. Again, this was very much accepted, and empathy and action were the responses to news that I'd need a sick day or starting leave early.

So if I can share one thing to think of as an employer, it's that your reactions and actions when an employee is at their most vulnerable really matter. It's those moments your employee will remember the most and think of how much they value you and the workplace culture you represent, or the exact opposite!

WHEN TO ANNOUNCE

Deciding when to announce pregnancy at work is a very personal decision. Some may tell everyone very early, while others may wish to keep the news to themselves for as long as possible. Those who elect to wait to advise their employer of their pregnancy may be concerned about the risk of miscarriage, which is much higher in the first trimester than in any other. Some may worry about discrimination. While these concerns are not unfounded, there can be many advantages to announcing pregnancy at work early instead of waiting.

Pregnancy loss is something that can unfortunately not be prevented entirely. The common practice of not announcing pregnancy until after the first trimester is centered around the goal of not letting others know about the pregnancy until it is less likely to be lost. This has the result of perpetuating the idea that pregnancy loss should not be talked about when it happens. Some may not wish to, but for those who would want support and community in the unlikely event that a pregnancy loss occurs, announcing a pregnancy early can be very helpful.

From a health and work safety standpoint, announcing pregnancy early has many advantages. Pregnant employees may need to modify their duties and/or avoid certain activities altogether for their own benefit and for the protection of their pregnancy. Although some of these changes might feasibly be made at the individual level, others will be more difficult to enact consistently without the support or intervention of management and colleagues. And because the first trimester is a time when an embryo is more sensitive to many workplace hazards, there can be great value in at least announcing a pregnancy to a supervisor or manager who can facilitate any needed changes or accommodations.

An employer cannot be expected to provide pregnancy-related job protection if they are not aware of the pregnancy. This means that pregnant individuals who miss work, show up late, or have changes in their performance due to common pregnancy conditions like fatigue and nausea might be disciplined by their employer and could even lose their job if their employer does not know that they are pregnant. The earlier a pregnancy is announced, the more time a pregnant person will have to become informed about the workplace benefits available to them and make plans for the rest of their pregnancy and parental leave.

BOX 8.2   Practice profile: Announcing my pregnancy at work

Jessica Nickless, DVM

My husband and I were in the midst of IVF when I moved from one job to another. My previous job was aware of our infertility issues but did not have a good work–life balance and had a lot of on-call duties that contributed to a high level of stress and mental health compromise. My new job was a corporate job, but with a private practice feel—they prioritized that balance and agreed to everything I asked for. Given our difficulties, I chose not to tell my new job we were trying to get pregnant before starting, especially since we would have only gone through one round of IVF after failing IUI at that point. But then, we got a positive pregnancy test. I felt guilt and worry. Would they resent me for not telling them that we were trying to conceive? Would they be upset that their new hire would end up on maternity leave after less than a year working there?

I ended up waiting until I was nine weeks pregnant and had a confirmation of a heartbeat on ultrasound, and then scheduled a meeting with management. I was very anxious—on top of my severe pregnancy nausea—but managed to blurt it out. I don't even remember what I said—I'm not sure it was anything more than "Hey, so I'm nine weeks pregnant"—I was just worried about their response. My medical director was immediately very happy for me and very supportive; my practice manager took a little longer as she was worried about the practice going back to a single-doctor practice, but she came up to me at the end of the day and did tell me how happy she was for me.

The practice manager immediately set changes in place to protect me (I wasn't to take radiographs; I wasn't to intubate or extubate). After discussing my history of infertility (which I felt comfortable enough to share—they did not ask or press), they suggested waiting until 12 weeks to tell the rest of the staff. As it turned out, my practice manager had gone through infertility herself and didn't want me to regret telling anyone too early. She reached out to higher management, and they helped me set up maternity leave (in Massachusetts, there is a maternity leave program through Paid Family and Medical Leave) and assured me to take however long I wanted, and my medical director even asked if I wanted to cut back my hours!

I ended up telling the rest of the staff at a staff meeting. They all became very protective of me, and toward the end of my pregnancy changed my appointments so that I did not see any giant breed dogs or aggressive animals. I was very thankful to have such a support system throughout my pregnancy. I ended up working until 39 weeks before starting medical leave, but my little one didn't make an appearance until 41+2!

HOW TO ANNOUNCE

How an individual communicates their pregnancy will depend on their usual method of communication with their supervisor and/or colleagues. If they normally interact face-to-face, scheduling an in-person meeting during a convenient time of day for both parties is appropriate. For situations in which face-to-face contact is not typical, a phone call or video meeting can work. In other cases, an email may be appropriate. Although typically a written announcement is not required, a pregnant employee may wish to at least document the conversation for their own records.

The first meeting may be very short, often only a statement that the individual is pregnant. If the estimated due date is known, this can be shared to give an employer a sense of when to expect the employee to be on maternity leave. An employer may be able to provide details on the company's maternity policies including job protection, benefits, and resources. They may also provide a summary of known workplace hazards and an expectation of the paperwork requirements for parental leave, paid leave, disability insurance payments, and so forth. If an employer has a radiation policy that pregnant employees will not be exposed to radiation, for example, this should be communicated to the employee. If this information is not offered, it is advisable to ask for it and to agree to meet again soon after to confirm that all needed information has been compiled and communicated.

If the pregnant individual is already aware of the accommodations they need, this can be expressed at the initial meeting. Some people, for example, experience severe morning sickness, fatigue, or other complications in the first trimester and may find that they need more breaks to sit or rest, make a change in their schedule and/or hours worked, or require flexibility for upcoming medical appointments. Other accommodations will depend on the individual's normal work responsibilities, such as restraining animals, handling hazardous drugs or chemicals, exposure to radiation and/or anesthesia, and so on. If an employer agrees to provide accommodations, it is advisable to formalize the agreement in writing to prevent any future confusion or conflict.

The employee may wish to communicate initial plans regarding the intended length of maternity leave, plans to return after leave, and thoughts on creating a different schedule and/or changing the number of working hours. The employer can advise the employee of available options and provide feedback on the employee's plans, all with the understanding that they will reconvene at a future time to confirm these plans or discuss any changes to them. If an employee chooses to reveal their pregnancy to only a small number of people, it should be made clear that it is not to be discussed publicly until the employee decides to disclose their pregnancy to others. Box 8.4 lists appropriate and inappropriate responses of an employer and/or employees to a pregnancy announcement or disclosure in the workplace.

BOX 8.3   Practice profile: Announcing pregnancy in vet school

Jessica Jaquin, DVM

A positive pregnancy test was revealed during my second year of veterinary school after I cried in Walmart over day-old flowers being tossed. During this time a pregnancy was never unwanted but not something on my radar. I took plan B the following morning, I was just a month into my engagement, I was a veterinary student in the middle of didactics, and my fiancé lived in Kansas. I remember during that time being completely overwhelmed with both joy and fear. I have always wanted to be a mom but at the same time I was on track to fulfill the career path I had dreamed about from five years of age. What was going to happen? I could hardly manage this vet school thing already, how was I going to navigate motherhood?

Announcing my pregnancy to my fiancé, Ben, was easy. I had his support, but he lived a state away while he completed his degree. Announcing my pregnancy to my classmates presented itself as a little more problematic for I didn't know how they would then perceive me from that time forward. It became announced when I could no longer stand the formaldehyde smell from the cadavers. I would vomit and just was adamant that I could no longer be the person going into the freezer to do that. One day a lab partner got so frustrated with me and I just snapped back, I am not walking in that freezer because I am pregnant! The rest of the class knew by the end of the day.

The most challenging announcement ended up being the administration meeting. I was nervous. Would I be let go? How could I even begin to think I could handle this? I left that meeting feeling empowered. I received nothing but support. I was given options—the amount of time I could delay exams, how to receive financial assistance, maternity leave times, and rejoining the class below me. Administration provided a list of support staff and contacts for previous students with similar life circumstances. I set plans to take a short maternity leave during Christmas break to avoid any disruption to my studies.

To this day I believe the stress of studying for pharmacology sent me into preterm labor. At 36 weeks, the morning of the exam, I dreamt I was having to cross an ocean to make it to the exam on time. I awoke to my water breaking prematurely. After a free ambulance ride for driving myself to the wrong university hospital, Ben was able to make it. I remember him redirecting me to the task at hand for I tried to study during labor. I was then induced and 19 hours later I got to hold my daughter, Harper. She was perfection. I remember as I was holding her for that first time, I could not let her go, she was all that mattered going forward. I had never felt anything like that moment. I never once imagined something could come before my aspirations as much as she did. I quickly realized after those first few days, that returning to take the 7+ exams I was behind on was unrealistic for me.

I did not have the support at home for any assistance with her newborn care and with her prematurity came some challenges in her feeding routine. For these reasons I submitted an email, administration took care of it, signed some papers, and I then joined the 2018 class. I think the hardest part was not knowing how the next year would look. I also felt connected to my 2017 classmates; it was hard being left behind.

The year I took off was astonishing for my mental health. I got to be a mom. I juggled motherhood and multiple part-time jobs. I worked for a horse barn and a daycare gym. I didn't realize six months into this break though that I would need to begin repaying back my student loans. This gave me the push to begin the master's in public health program. This program was online and overlapped with the veterinary curriculum quite a bit. I survived on these part-time job incomes, student loans, WIC, and food stamps.

When I returned to classes that following year, I did well on that pharmacology exam and started to grasp motherhood. That year gave me the ability to ease back into classwork, juggle motherhood, and rediscover myself. I felt like when I returned my grades and study habits improved. I had to be focused during lectures. I had limited time to study so when I did, I was much more productive in retaining the material. Now I don't think I would recommend this study habit to everyone, but it did work for me. I went off to get married and had a second child during school. I had a son, Tristan, during a free block of clinics. During clinical rotation it was easier to move around, and I was able to provide myself with a 12-week maternity leave. I never had trouble with making my OBGYN appointments or accommodations.

Overall, I felt the school was very accommodating during both of my pregnancies and I had a great experience at the University of Missouri. I would offer the advice of not being afraid to reach out for the support of your classmates and administrators. Hopefully just like me, they will surprise you with overwhelming support.

BOX 8.4 Receiving a pregnancy announcement: A guide for employers and coworkers

APPROPRIATE COMMENTS AND/OR REACTIONS TO LEARNING AN EMPLOYEE (OR COWORKER) IS PREGNANT

- Offer congratulations;
- Ask if the employee has any immediate needs;
- Offer support and encouragement;
- Present workplace protections and policies, discuss performing a risk assessment, reference available maternity benefits, FMLA, short-term disability coverage;
- Direct the employee to Human Resources or another department as necessary;
- Offer to connect the employee with other employees who are parents or who have navigated pregnancy, maternity leave, and/or returning to work;
- Make plans to meet again within two weeks to check in with the employee and receive feedback and/or change plans if needed.

INAPPROPRIATE COMMENTS AND/OR REACTIONS TO LEARNING AN EMPLOYEE (OR COWORKER) IS PREGNANT

- Asking if the employee is pregnant before they have announced it;
- Expressing anger/frustration/disappointment regarding how the pregnancy and/or maternity leave will negatively affect the practice revenue/workload/schedule;
- Asking any questions about family planning, infertility, or if the pregnancy was intended;
- Commenting on the employee's body or behavior;
- Commenting on any gossip surrounding the employee;
- Sharing the employee's pregnancy with other personnel without explicit permission to do so;

- Making any threats or statements regarding dismissal, loss of opportunities, or loss of pay;
- Making any negative comments about the employee or their future at the company;
- Expressing an unwillingness to discuss accommodations, benefits, maternity leave, and other resources.

TAKE-HOME POINTS

- In many countries, pregnancy and its associated conditions are protected by law from discrimination in the workplace. Most employers understand their obligation to protect and support pregnant employees and parents in the workplace;
- There is room for improvement in terms of pregnancy protection laws in some countries, and a need for more consistent enforcement of existing laws across all countries;
- Unfortunately, there are still cases where discrimination occurs despite laws being in place to prevent it. Not all cases are reported because of fear of retaliation. However, reporting and holding accountable employers who violate the law is the best way to improve the experience for future employees;
- Employers who are not respectful of employee pregnancy in the workplace will likely not be respectful of the needs of working parents and families;
- Announcing pregnancy early to a manager or other key individual(s) is the best way to provide protection from workplace hazards, both to the individual and the pregnancy. It can also give employees helpful time to plan for maternity leave and beyond;
- In most countries, employers are required or encouraged to provide accommodations to pregnant and/or breastfeeding employees who request them as long as they do not cause undue hardship for the rest of the practice's employees. Employers are generally not supposed to require accommodations or assume that pregnant employees need them, so it is often up to the employee to make the request. Employees cannot be discriminated against for requesting an accommodation;
- Individuals can file a claim with the governing body in their country if they feel they have experienced discrimination. Most countries have laws preventing retaliation by employers after a claim is made;
- Most employers and/or coworkers will be happy and supportive when a pregnancy is announced. However, some commonsense guidelines can help these individuals understand appropriate and inappropriate ways to respond.

REFERENCES

1. US Equal Employment Opportunity Commission. Legal rights of pregnant workers under federal law. Accessed October 19, 2022. https://www.eeoc.gov/laws/guidance/legal-rights-pregnant-workers-under-federal-law

2. US Equal Employment Opportunity Commission. Pregnancy discrimination and pregnancy-related disability discrimination. Accessed October 19, 2022. https://www.eeoc.gov/pregnancy-discrimination

3. American Association of University Women. 7 things to know about pregnancy discrimination. Accessed October 19, 2022. https://www.aauw.org/resources/legal/7-things-pregnancy-discrimination/

4. US Department of Labor. Family and Medical Leave Act. Accessed November 5, 2022. https://www.dol.gov/agencies/whd/fmla

5. US Department of Labor. Women's Bureau. Accessed November 2, 2022. https://www.dol.gov/agencies/wb

6. Canadian Human Rights Commission. Policy on pregnancy and human rights in the workplace - Page 1. Accessed November 2, 2022. https://www.chrc-ccdp.gc.ca/en/resources/policy-pregnancy-human-rights-the-workplace-page-1

7. United Kingdom Government. Pregnant employees' rights. Accessed November 2, 2022. https://www.gov.uk/working-when-pregnant-your-rights

8. Advisory, Conciliation and Arbitration Service. Managing your employee's maternity leave and pay. Accessed November 2, 2022. https://www.acas.org.uk/managing-your-employees-maternity-leave-and-pay/discrimination-because-of-pregnancy-and-maternity

9. Australian Government. Pregnant employee entitlements. Accessed November 2, 2022. https://www.fairwork.gov.au/leave/maternity-and-parental-leave/pregnant-employee-entitlements

10. Addati L, Cassirer N, Gilchrist K. *Maternity and Paternity at Work: Law and Practice Across the World.* Geneva: International Labour Organization; 2014.

11. Mykhalchenko O, Santagostino Recavarren I. In 38 countries, women can still be fired for being pregnant. May 13, 2021. Accessed November 2, 2022. https://blogs.worldbank.org/developmenttalk/38-countries-women-can-still-be-fired-being-pregnant

12. Schulte B. What's fair treatment for pregnant workers? The U.S. isn't sure. Other countries are. December 5, 2014. Accessed October 19, 2022. https://www.washingtonpost.com/blogs/local/wp/2014/12/05/whats-fair-treatment-for-pregnant-workers-the-u-s-isnt-sure-other-countries-are/

13. U.S. Equal Opportunity Commission. What you should know about the pregnant Workers Fairness Act. Accessed March 31, 2023. https://www.eeoc.gov/wysk/what-you-should-know-about-pregnant-workers-fairness-act

14. ACAS. ACAS working for everyone. Accessed April 1, 2023. https://www.acas.org.uk/about-us

15. Australian Human Rights Commission. Complaints. Accessed April 1, 2023. https://humanrights.gov.au/complaints

16. Batha M. Major UK firms act to end "discrimination scandal" for new mums. September 27, 2018. Accessed April 1, 2023. https://www.reuters.com/article/us-britain-employment-women/major-uk-firms-act-to-end-discrimination-scandal-for-new-mums-idUSKCN1M71SS

9

"So, how long are you going to be on vacation for, doc?"

Planning parental leave

During each of my pregnancies, I felt a great deal of anxiety and confusion about maternity leave and what I was entitled to. I worried about meeting the federal requirements for job-protected leave, either because of how long I had been working for my employer or because of how small my private practice was. In all four cases, I was able to take at least 12 weeks of leave (I took a longer unpaid leave of absence when my second child was born). In each case as I planned my leave, I had to immediately start worrying about how our family would cover our expenses without my income. We had to start saving aggressively to cover the time when I would receive no paycheck. Even when I worked for a corporate practice, there was confusion about whether I was entitled to paid time off and whether taking it would put me in a production deficit that I would have to pay back. There was no clear policy regarding parental leave of any kind. My husband also had no paternity leave benefits offered to him. The anxiety and stress caused by these experiences made me feel very unsupported as a veterinary professional and as a mother.

DOI: 10.1201/9781003406907-9

Parental leave can be a stressful thing to plan for veterinary professionals. First, there are the unknowns surrounding when the baby will be born and whether any health complications will necessitate an earlier start to maternity leave. Then there are the concerns about how much leave is allowed, how much is paid during the leave, and how much leave is affordable. In addition, it can be hard to resist the urge to worry about how clients, patients, coworkers, and/or employees will cope with the absence of a valuable member of the team. For non-birthing parents, there is often also the stigma associated with taking anything more than a few days off after the baby's arrival. All of these factors can create anxiety and other challenges when trying to plan their leave. However, taking the time—to heal, to bond, to support a partner, to establish routines—can be hugely beneficial for many families and for the success of the working parent(s) when they return to work. This chapter reviews the benefits of parental leave for all parents; the varying laws and policies regarding parental leave and paid parental leave by state, country, and company; and the strategies that can help the planning of parental leave to be more successful and less stressful.

The term "parental leave" can mean multiple things. In some cases, it is used as a gender-neutral term that includes leave by either parent. In some countries, Mercer reports, parental leave specifically refers to "additional leave that is typically taken in the first year of the child—or sometimes up to age three" meant to care for a baby or young child.[1] Some companies use the term *parental bonding leave* to refer to benefits provided to adoptive parents, parents who used a gestational carrier, and LGTBQ couples and others who do not identify with the terms "maternity" or "paternity."[2] For the purposes of this book, "parental leave" will be used as a general, gender-neutral term unless otherwise specified. Maternity and paternity leave often are assigned very different benefit levels and have different stigmas attached to them. Thus, they will often be referred to separately to highlight the current state of each while comparing them to each other and across different countries, states, and companies.

BENEFITS OF PARENTAL LEAVE

MATERNITY LEAVE

Contrary to what some either mistakenly or jokingly suggest, maternity leave is not a break or a vacation. It is a very necessary period to allow for the healing of the body after childbirth (whether vaginal or cesarean delivery, routine or complicated) for those who have given birth. It allows for the care of a very vulnerable newborn who requires round-the-clock attention and for the protection of this newborn from exposure to contagious diseases. Parental leave is also filled with frequent doctors' visits and bonding between parent(s) and baby and family before one or both parents return to work. In some cases, it will include care of an older infant or toddler, or bonding with a child after foster placement or adoption.

The National Partnership for Women and Families reports that paid maternity leave helps to improve the physical and mental health of mothers and infants. Birthing parents who are given paid leave are less likely to report intimate partner violence. Infant mortality rates are lower as well, and fewer children under the age of two years are subject to head trauma (such as from being shaken) at the hands of stressed parents. The length of the paid leave was also found to be significant. For example, parents who had less than eight weeks of paid maternity leave had poorer overall health and increased rates of depression. Each additional week of paid leave lowered the risk of poor mental health by 2%. Longer paid leave also increased the likelihood of continued breastfeeding, which can have many health benefits for both mother and baby. Paid leave that lasted more than 12 weeks also resulted in higher rates of infant vaccination.[3]

Compared with mothers who took no leave or had completely unpaid leave, mothers who took paid leave had a 51% lower chance of needing to be re-hospitalized, and the rate of re-hospitalization of infants was reduced by 47%. Those with paid maternity leave were almost two times as successful at stress management and participating in regular exercise. States that began mandating paid leave saw fewer low-birthweight and preterm births, particularly for Black mothers. Unfortunately, paid leave is significantly less likely to be offered to workers paid a lower wage, who are disproportionately people of color. Therefore, having access to paid leave can help reduce racial and ethnic disparities that affect health, well-being, mortality, and safety of mothers and babies.[3]

PATERNITY LEAVE

While it still lags far behind maternity leave in many countries, paternity leave also creates significant benefits for not only fathers (and other non-birthing parents), but all family members. These benefits include increased bonding, more equitable distribution of household labor, increased overall family income, and reduction in the gender wage gap. The effects of paternal involvement in childcare and paternity leave have also been shown to be long-lasting, affecting the distribution of childcare and work in and outside of the home in the future.

Fathers who took paternity leave reported developing a "special" bond with their child(ren) that had long-lasting effects. They were able to better prioritize their relationship with their child, participate in their child's daily routine, and feel more satisfied that they were there for their children. This effect of increased "engagement in developmental tasks and caretaking" after taking paternity leave has been shown to continue throughout the first few years of a child's life, long after paternity leave has ended.[2] In addition, in a study of 126 fathers who took paternity leave, 100% of them reported being happy with their decision to take leave.[2] Sixty percent of men in one study described their time taking care of their children as "very meaningful," which was almost twice the percentage who reported the same feeling about their paid employment.[4] Many fathers who took paternity leave also reported having increased appreciation for their employers.[2]

In addition to the benefits to father and child, 90% of fathers who took paternity leave reported an improved relationship with their partner. "That may be because when fathers take leave, it signals a greater investment in family life," Colantuoni et al. suggest, "reducing the burden on the other parent and strengthening parental relationships." This can help to reduce the risk of postpartum depression in the partners of those who take paternity leave.[2] Hald Anderson suggests that increased uptake of paternity leave relative to the maternity leave of their partner allows fathers to increase their "household capital," or the amount of work they perform at home, which allows for a narrowing of the gender wage gap. As fathers perform a more equitable share of child rearing and other tasks, mothers can feel more empowered to pursue work outside of the home and may be able to work more hours and/or pursue career advancements that require more of a time commitment and/or travel. These opportunities may increase their earning potential and help it to be more in line with that of their partner's earnings.[5]

Unfortunately, the availability of paternity leave and the willingness of employees to utilize what is offered both still have great room for improvement. As of 2021, about 90 of 187 countries surveyed require paid paternity leave, with 38% of organizations in those countries exceeding the statutory minimum. Even when paternity leave is offered, however, fewer than half of the eligible fathers take the full amount of leave that is available.[2] Encouraging increased access to and acceptance of paid paternity leave can benefit both parents. Mercer explains that "by creating policies that include both men and women, we are able to identify and eliminate barriers to taking leave for everyone."[1] Fathers who ask for and take advantage of paternity leave create an increased demand and attention on these types of benefits, which will allow more mothers to have access to them and feel empowered to take them as well.

PARENTAL LEAVE LAWS AND POLICIES

PARENTAL LEAVE BY COUNTRY

According to the Organisation of Economic Co-operation and Development (OECD), the United States is the only one of the 38 member countries that does not offer at least 12 weeks of paid maternity leave nationwide.[6] The minimum length of the leave, the percentage of regular pay that is covered, and the option to take additional paid time vary significantly by country. In addition, the ways in which paid leave is financed vary from country to country. The Bipartisan Policy Center reports that many OECD countries use "social insurance funds that are supported by employer, worker, and government contributions" to finance paid leave. In some countries, taxes, health insurance, or public long-term care insurance are utilized to pay for paid parental leave.[7] Table 9.1 includes a comparison of the lengths of maternity leave and percentage of pay guaranteed to working mothers (referred to as primary caregivers in some countries) by country.[8]

Paternity leave, paid or unpaid, is still much less common and less robust than maternity leave in most countries. The Bipartisan Policy Center states that 27 of 38 OECD countries offer some paid paternity leave to fathers (sometimes erroneously referred to by some entities as "secondary caregivers") that typically includes job protection. Twenty-three of 38 OECD countries offer paid parental leave, which is useable by either parent. To encourage more fathers to take advantage of these benefits, some countries have restrictions on how much of the time can be used by fathers only ("daddy quotas"). Some countries will incentivize fathers taking paternity leave by offering additional weeks of paid leave to couples in which both parents use the leave they are offered (bonus months).[9,10] Table 9.2 summarizes paternity leave protections and benefits offered by a selection of countries.[11]

Table 9.1 Maternity leave offered by country

Country	Minimum length of job-protected leave (wk)	Percent of salary paid (%)	Optional additional time at percent paid
Bulgaria	58.6	90	51.9 wk at 28%
Greece	43	61.8	0
United Kingdom	39	29.8	0
Slovakia	34	75	130 wk at 34%
Croatia	30	100	26 wk at 64.5%
Czech Republic	28	59.4	47.1 wk at 82.3%
Ireland	26	27.3	2 wk at 27.3%
Hungary	24	70	136 wk at 36%
New Zealand	22	47.5	0
Australia	18	42.4	0
Chile	18	100	12 wk
Colombia	18	100	0
Finland*	17	74.8	32 wk at 19%
Canada	16	47.7	35 wk at 52%
Japan	14	67	44 wk at 59.9%
Norway	18	96	68 wk at 33.4%
Iceland	17	77.9	8.7 wk at 78%
France	16	95.7	26 wk at 14%
Mexico	12	100	0
United States	0–12	Unpaid, unless sick leave or vacation time available. Must be eligible for FMLA to qualify. Some states/employers offer additional benefits	State and employer dependent (none at federal level)

Table 9.2 Paternity leave offered by country

Country	Minimum length of job-protected leave (wk)	Percent of salary paid (%)	Optional addition at percent paid
Finland[27]*	17	74.8	32 wk at 19%
Australia	2	42.4	0
Austria	4	18	9 wk at 50.4%
Belgium	2	75	17.3 wk at 21%
Bulgaria	2	90	0

(Continued)

Table 9.2 (Continued)

Country	Minimum length of job-protected leave (wk)	Percent of salary paid (%)	Optional addition at percent paid
Canada	0	Unpaid	5 wk unpaid
Czech Republic	1	59	0
Ireland	2	27	2 wk unpaid
Hungary	1	100	0
New Zealand	0	Unpaid	0
Chile	1	100	0
Colombia	1	100	0
France	2	96	26 wk at 14%
Mexico	1	100	0
United States	0–12	Unpaid, unless sick leave or vacation time available. Must be eligible for FMLA to qualify. Some states/employers offer additional benefits	State and employer dependent (none at federal level)

*As of 2022, Finland offers 320 working days of paid parental leave. In families with two working parents, each parent may take 160 working days of paid leave. In addition, one parent may give up to 63 of their eligible paid working days to the other parent if desired. A single parent may take the full 320 working days of paid leave.

BOX 9.1   Practice profile: Maternity leave and cultural expectations

Sam Cutts, MA, VetMB, MRCVS

AS TOLD TO EMILY SINGLER

Sam Cutts is an equine veterinarian and practice owner at a large practice in England that treats small animals, small farm animals, and horses. She is 1 of 6 partners who own the practice of 100 employees. In contrast to the United States, in England it is often expected for mothers to take a year of maternity leave. Sam reports that this cultural expectation made it easy to plan to take the amount of time she needed to heal and be home with her new baby. However, the expectation to take so much time can result in mothers who choose to take less time feeling guilty and judged for their decisions.

After her first child was born, Sam planned to take two months of leave and then slowly phase back in to working. She chose not to take a longer leave because she was a practice owner and needed to be involved in the management of it. Once she was on leave, she also felt that her mental health was negatively affected both by the challenges of staying home and the isolation caused by the COVID-19 pandemic. When she did transition back to work, she had a maternity nurse (at her own expense) who would bring her baby to her workplace to breastfeed and also help care for the baby at night as well so Sam could sleep. For the first year after she came back, she was not on call since she was breastfeeding.

At the time of this writing, Sam was on her second maternity leave, and she planned to take 3 months with a phased return to work. She has seven other equine associates who can cover the load while she is on leave, and she still meets with her partners to stay involved in the management of the practice.

When asked about the challenges she faces, Sam points out that all her partners are male. She feels the need to bring the perspective of a working mother into her partnerships. She also feels the need to compete with corporate veterinary practices that offer six months of fully paid maternity leave and a return-to-work bonus. She feels concern about putting more pressure on the rest of the team when she is on leave, even though everyone is supportive. She feels pressure from herself to work during leave, especially as a business owner.

One of the most unhelpful aspects of maternity leave in the UK is the assumption that every mother will take 12 months. While in the US there is pressure for parents to not take too long of a leave, Sam feels the reverse is true in her country. She says it is very uncommon for mothers to return to work sooner than 9 months due to the social pressure to take a leave of that length. In some cases, people will take a longer leave than they can really afford (since their maternity leave pay is often significantly less than their regular pay) to avoid being judged for their decision to go back to work "too soon." Sam also knows that taking such a long leave would not be good for her mental health, and states that it felt good to get back to work when she did.

When asked what advice she would give to other working mothers, Sam stresses that each individual should decide what is right for them. "Don't let society decide for you," she counsels. Sam is also a big proponent of a phased return. The example she gave was to start working partial days for the first month and then gradually increase the hours worked over the next 2–3 months. Every case is different, however, so it helps when working parents feel the freedom to discuss their needs on an individual basis with their workplace. It can be hard to avoid putting pressure on yourself and letting cultural expectations determine what it means to be a good mother. "Whatever makes you happy," Sam counters, "makes you a good mum."

FAMILY AND PARENTAL LEAVE IN THE UNITED STATES

The National Partnership for Women and Families reports that the United States is "one of just a few countries in the world with no national paid leave of any kind." As a result, more than 80% of working parents do not have access to any paid leave after the birth or adoption of their child. Furthermore, 46% of workers are not even eligible for unpaid job protection.[3]

According to the National Conference of State Legislatures (NCSL), the only nation-wide parental or family leave protection in the United States is the Family Medical Leave Act (FMLA). This provides "up to 12 weeks of unpaid leave during a 12-month period to care for a newborn, adopted or foster child, or to care for a family member, or to attend to the employee's own serious medical health condition." In order to qualify, an employee must work at a business that has more than 50 employees, and they must have worked at the business for at least 1250 hours and at least 12 months. Because this leave is not required to be paid, it merely serves as job protection for the duration of the leave. If the employee has available paid sick, personal, or vacation time, they may use this to receive pay during some or all of their leave.[12,13]

Some states have passed laws providing additional protections and benefits to their own residents. The NCSL reports that 11 states offer some form of paid family leave, all funded through employee-paid (and sometimes employer-paid) payroll taxes. These states are California, Colorado, Connecticut, Delaware, Massachusetts, Maryland, New Jersey, New York, Oregon, Rhode Island, and Washington State. The District of Columbia also offers paid family and medical leave. In addition, 16 states and the District of Columbia also require paid sick leave. Three states offer paid parental leave for state employees, which include the birth, adoption, or fostering of a child. Each state has its own criteria in terms of eligible employers (based on the size of the company) and employees (based on how long they have worked at the company). Table 9.3 summarizes the leave benefits offered by states that offer additional protections apart from the FMLA.[12]

BOX 9.2 Practice profile: Planning my maternity leave, 2020

Anne Budzinski, DVM

As the first person to be pregnant at the exclusively cattle practice where I am one of three associates, it was a unique endeavor to plan my maternity leave. We did not have an employee handbook or protocol in place yet, so it was a discussion where I met with the owner, the scheduler, and the accountant to lay out a plan about 2 months prior to my due date.

Since there wasn't anything in place, I requested 6 weeks off, and I would use my remaining paid time off and sick time to offset that stretch of time away from the clinic. I had 2 weeks left to use, so I was thinking I would have 4 weeks unpaid. It was determined that 6 weeks was too long for me to be away from the practice, and it was granted that I could have 4 weeks of maternity leave. I believe this was put in writing somewhere after our meeting though I don't recall it at this time.

A few months later we received an employee handbook that stated parental leave is allocated to 4 weeks paid, and we have the option to use personal days to equal a total of 6 weeks off. I worked up to my due date, was induced, and had my son via emergency cesarean section. I had 4 weeks home with him and then went back to work, and I was at full hours at 6 weeks postpartum. Looking back now I wish I had urged for more time off—more like 6–8 weeks in order to recover from surgery and also spend time with my son.

EMPLOYER-OFFERED PAID LEAVE

Employer-paid parental leave is slowly becoming more common across the United States. As of 2021, about 55% of employers in the United States offer some type of paid maternity

Table 9.3 Parental leave offered by US state

State	Leave offered
California	Up to 12 weeks unpaid leave plus 4 months of maternal disability (paid); qualifying employees can also receive up to 8 weeks of paid leave at 55% of regular pay
Colorado	Up to 12 weeks of paid leave for qualifying employees
Connecticut	Up to 12 weeks every 2 years of paid leave for qualifying employees
District of Columbia	Up to 8 weeks of paid bonding time
Delaware	Up to 12 weeks of paid leave for qualifying employees
Hawaii	Up to 4 weeks of unpaid leave per year for qualifying employees
Maine	Up to 10 weeks every 2 years of paid leave for qualifying employees
Massachusetts	Up to 12 weeks of paid leave for qualifying employees
Minnesota	Up to 12 weeks of unpaid leave for qualifying employees
New Hampshire	Up to 6 weeks of paid leave (at 60%) for employees of businesses who choose to participate
New Jersey	Up to 12 weeks of unpaid leave for qualifying employees
New York	Up to 12 weeks of paid leave for qualifying employees
Oregon	Up to 12 weeks of paid leave for qualifying employees
Rhode Island	Up to 5 weeks of paid leave for all employees
Vermont	Up to 12 weeks of unpaid leave for qualifying employees
Washington	Up to 12 weeks of paid leave for qualifying employees
Wisconsin	Up to 6 weeks of paid leave for qualifying employees

leave.[14] This does not mean, however, that every employee will be eligible for these benefits. Many companies have tenure requirements, such that employees who have not worked 12 full months with the company are not eligible for paid parental leave. Others are disqualified because of the number of hours they work, because they were hired as independent contractors, or because they are paid hourly. This can mean that higher-paid employees will reap the benefits and lower-paid employees will not. Some companies require their employees to be designated as either a "primary caretaker" or a "secondary caretaker" of their child, and they provide different benefit levels to each designation. Some employees will choose not to use the whole benefit they are offered out of concern for the effect their leave will have on their career, their clients, and/or their colleagues.

At least 180 US companies offer their eligible employees at least 16 weeks of paid parental leave, and some offer as much as 52 weeks of paid leave.[14] Although there is no national required paid parental leave in the United States, the Department of Commerce reports that the Federal Employee Paid Leave Act (FEPLA) "makes paid parental leave available to certain categories of Federal civilian employees." This act provides up to 12 weeks of paid parental leave to eligible employees after the birth or adoption/foster placement of a child.[15]

Paid parental leave is still not nearly as common or as comprehensive in clinical veterinary practice when compared with other industries. Some veterinary corporations do advertise that they offer anywhere from 8 to 16 weeks of paid parental leave to their full-time employees. In addition, some privately owned practices may offer various amounts of paid or partially paid leave.[16–20] For employees who are paid on production with negative accrual, drawing a paycheck during their maternity leave might result in a deficit that they have to make up upon their return to work.

BOX 9.3   Practice profile: A tale of two maternity leaves

Yao A. Yao, DVM

Dr Yao is a mother of two and currently working in industry. Before that, she spent seven years in small animal emergency medicine. While she loved emergency medicine, she struggled with the inconsistency and timing of her shifts (working overnights, holidays, and weekends). When she began her first maternity leave shortly before her first child was born, she still thought she would return to emergency work at the end of her leave. Her work offered a generous 6 months of mostly unpaid, job-protected leave. She was able to use some short-term disability coverage for the first 6 weeks of leave, and she had some paid time off accumulated as well, but otherwise her leave was unpaid. After the birth of her baby, however, she decided not to continue in emergency work due to the inconsistency of her schedule that would have required her to miss many evenings, weekends, and holidays with her family. She instead did some general practice relief work for a while.

She was able to secure a completely remote position working for a pharmaceutical company, that she refers to as a "unicorn job." Although she took a pay cut compared with emergency work, she describes the quality-of-life improvement as more than worth it. When she announced that she was pregnant in her industry job, she recalls how different her experience was. Her manager introduced her to other employees who had been on maternity leave recently. She was given documents detailing all of the company's policies relating to pregnancy and maternity leave. Her manager just asked her when she would go on leave and when she wanted to come back. She again received 6 weeks of short-term disability insurance pay (full pay!), but her company also provided an additional 12 weeks of fully paid leave and then provided additional unpaid leave.

At the time we spoke, she was on maternity leave. Despite not getting much sleep, she seemed happy to report that she felt no pressure to go back to work, that she was being totally left alone while on leave, and that her manager had even reprimanded her for logging in to her work email to check her messages. Her work allows her the flexibility to "see how it goes" in terms of when she decides to go back, and other workers have set the precedent of taking 6 months of leave, and that is considered to be common in her department.

Her advice for other moms and soon-to-be moms is to remember you are not stuck. She remembers falling into the mindset of "that's just what you do" and feeling that she had to struggle through even when her work schedule and environment were not contributing to a good quality of life. She wants other moms to know that there are other opportunities out there for veterinarians who are not happy in their current role or who do not have the work–life balance that they need. A veterinary degree and experience can also open doors into related fields in human medicine and beyond. She is happy that she has found a role that better supports her both professionally and personally, and she encourages other moms who need a change to do the same.

SHORT-TERM DISABILITY INSURANCE

Short-term disability insurance can help to partially cover maternity leave for employees who have given birth, but there are caveats to its use as well. It is meant to cover the period of disability associated with temporary conditions, including complications associated with pregnancy and childbirth and recovery from routine childbirth. Some employers offer group short-term disability coverage, and some individual policies may be available as well. In many cases, however, individual short-term disability coverage may not include coverage for recovery from routine childbirth (although it may still cover complications). Some disability insurance policies will have stipulations that employees who are already pregnant at the time they sign up are not eligible and/or that they must have the policy in place for a certain amount of time before they become pregnant to be eligible for benefits.[21]

Even when short-term disability coverage does cover childbirth, there can be exclusions. Many disability insurance policies will have an *elimination period*. This is the period of days or weeks during the disability in which no payment will be made. It can be considered a waiting period, where an employee could use paid vacation or sick leave if they have it, until the disability coverage begins. Short-term disability coverage also has limits in terms of how long benefits can last. This is called the *benefit period*. This will vary from policy to policy, but an average benefit period is usually three months. The period over which payments will continue will also depend on how long the employee is considered to be disabled. Although it is not uncommon for mothers to want to take 12 weeks or more not only to recover but to bond with their child, they may only be considered disabled and eligible for insurance coverage for 6 weeks after a routine vaginal delivery (longer for a cesarean delivery). The *benefit amount* specifies the percentage of regular income to be paid out by the insurance. Although it is sometimes close to 100% of an employee's regular pay, it is more often much lower (50–60%, for example).[21]

SELF-FUNDED LEAVE

In the many situations in which paid leave and/or disability insurance are not sufficient to fund parental leave, expectant parents are often left to save up enough money to cover their expenses while they are not receiving a paycheck. This can be most easily accomplished when it is started early in a pregnancy or adoption process. When determining the amount of money needed, it will be important to consider the cost of any health insurance premiums that will be due during the period of leave. These may need to be paid directly by the employee during leave (or before leave), or the employer may seek reimbursement after leave has ended. Registries now exist to allow individuals to crowdfund their parental leave savings with donations from friends and family.

PLANNING PARENTAL LEAVE

Beacom and Campbell describe a "thoughtful three-phase parental leave action plan" that includes a preparing phase before the birth of the baby, the parental leave phase itself, and the return-to work phase. Having these phases planned out ahead of time (to the extent possible) can help to reduce an expecting veterinary professional's stress level and provide reassurance that their needs (both at home and at work) have been communicated.[22]

It is ideal to plan for parental leave with as much anticipation as possible for a variety of reasons:

- To allow time to investigate the types of paid and/or unpaid parental leave available;
- To allow a non-birthing partner to determine if they also qualify for parental leave and plan for that;
- To complete appropriate paperwork for FMLA, short-term disability insurance coverage, and any employer and/or governmental paid leave that is available;
- To allow sufficient time to save up money to cover any unpaid portion of parental leave and to determine how much leave can reasonably be taken;

- To allow the workplace to make necessary staffing and scheduling arrangements, including hiring relief coverage.

In some cases, maternity leave starts shortly or even significantly before the birth of the baby. In some countries, this is standard to allow the pregnant individual to rest and to account for the possibility of the baby coming early. In some cases, an early start to maternity leave may be taken out of concern for the baby's or the mother's health. Complications such as hypertension, bleeding, incompetent cervix, and others can require the mother to be on bed rest or reduced activity at home or in the hospital for monitoring. Likewise, changes in the baby's growth, movement, or vital signs can warrant a recommendation to stop working earlier than might have otherwise been planned. Some people stop working for reasons such as back pain, swelling, and exhaustion, which can create challenges to continuing to perform a very physically demanding job.

PREPARATION AND PLANNING OF PARENTAL LEAVE

There are important steps that the pregnant individual can take to ensure that a satisfactory plan can be devised:

- Review the company's parental leave policy, including minimum and maximum length, paid leave, and any other benefits such as gradual reentry;
- Determine whether there are other leave protections and benefits offered by the country, state, or province;
- Review any short-term disability coverage that may be in effect;
- Determine how much leave is desired (this may change, but it is good to have an initial idea);
- Calculate any deficit between pay during leave and regular pay (or the amount of pay needed to cover expenses); be sure to include any insurance premiums or regular paycheck deductions that may still be due during parental leave (particularly in the US);
- Evaluate savings to determine if they can cover any deficit;
- Create a savings plan to cover any remaining deficit if possible;
- Approach the appropriate individual or department in the workplace to discuss a parental leave plan; try to plan a time for a meeting when both individuals can avoid being disturbed if possible;
- State the date when leave is anticipated to begin; also state a tentative return to work date;
- Verify any available paid leave benefits and determine how and when the payment will be made; this can include available vacation, sick, and personal leave days;
- For veterinarians paid on production, clarify whether negative accrual will be in force with any paid leave that might create a production deficit;
- Make plans for payment of any insurance premiums or other paycheck deductions that will be due during leave;
- Discuss desired level of communication during leave. For owners and those in managerial positions, there may be a need and/or desire to be kept in the loop on certain projects. Veterinarians and other staff should not be responsible for managing cases

or communicating with clients during their leave. However, some expecting parents may wish to be included in communications and kept up to date on any changes that happen in their absence;

- Discuss how ongoing projects and responsibilities can be transferred to other individuals or put on hold during maternity leave. This may include hiring relief/locum veterinarians or staff members, training another individual already on the team to carry out certain responsibilities, and/or preparing staff and clients for a disruption in certain services or projects during the upcoming parental leave;

- Discuss any available gradual reentry programs offered. This can be particularly valuable after a longer leave, but all individuals can benefit from a gradual reentry program;

- Discuss if changes in hours and/or schedule are anticipated after leave. This is also a good time to discuss accommodations for lactation, including pumping. Accommodations should include both a private space (that is not a bathroom) for pumping, refrigerator space (in a refrigerator not used for vaccines and other medications) to store expressed breastmilk, and the understanding that time will need to be blocked out in the schedule for the individual to express breastmilk, possibly as often as every 2–3 hours.

PARENTAL LEAVE FOR EMPLOYEES OF SMALL BUSINESSES

Although it may feel impossible for small businesses (which includes almost all noncorporate veterinary practices) in the United States to offer paid parental leave without the extensive financial resources available to larger companies and governments, it is possible for small businesses to offer paid leave programs. Doing so can be a huge recruiting and retention tool for business owners trying to attract and keep employees looking to grow their families. Would-be employees also often look for paid parental leave benefits as they search for employment.[9]

When deciding what kind of paid leave to offer, small business owners should do a cost-benefit analysis. Offering paid leave can be costly, but it can prevent some employees from quitting out of frustration, resentment, or disappointment because of a lack of workplace support. The costs associated with the loss of an employee can include lost revenue, increased recruiting costs, and sometimes the loss of additional employees who decide to follow the first employee who leaves. Opting not to offer this kind of benefit can negatively affect employees, even if they don't leave the practice. This can be noted as low morale, lower productivity, and less engagement in the work they perform. Finally, not boasting this type of benefit to employees can negatively affect the reputation of the company to the general public, who increasingly values companies that act with integrity and compassion toward not only their customers, but also their employees. Given these potentially significant business losses associated with no paid leave, it can make better financial sense to offer paid leave.[9]

It is important to talk with employees for their feedback on the types of benefits they would most appreciate, and to be sure to include all employees, not just women, in the conversation and in the paid leave policy.[9] Parentaly, a company that offers steps to improve parental leave for both employees and employers, outlines some flexible ways to accomplish paid parental leave in a small business climate.[10] These include:

- Researching ways to subsidize paid leave through state government programs, short-term disability insurance, and other programs;
- Offering a combination of paid and unpaid leave, or partially paid leave. This can look different for every business;
- Offering part-time work for full-time pay when employees first return;
- Hiring temporary help (relief/locum employees) to cover parental leave;
- Allowing parents to split their parental leave into smaller blocks of time (i.e., they go on leave, come back to work, then take another block of leave at a later time);
- Discussing ahead of time how parental leave will affect any bonuses or production pay;
- Planning to transition some projects to other employees and putting some projects on hold during parental leave.[10]

It can also be helpful to discuss ahead of time how employees will utilize any vacation, personal, and sick days they have at their disposal, and to make sure they know how much they have accumulated. Some employees may choose to save some of these days to prepare for the inevitable childhood illnesses that will require calling out of work to care for them. If any return-to-work programs are offered, this should be discussed before leave so the employee knows what to expect.

PARENTAL LEAVE FOR BUSINESS OWNERS AND SOLO PRACTITIONERS

Business owners, whether they work alone or have employees, don't always have access to the types of benefits employees may be eligible for. When they have children of their own, especially if they are solo practitioners, they may find they need to be more creative in finding the time to take for healing and/or bonding after their child arrives. It may feel impossible to take even six weeks of leave. Although every situation differs, there are some techniques that can help business owners to maximize their bonding and healing time and minimize their stress, without their business falling apart. They include:

- Setting boundaries ahead of time. This can reduce stress and help others know what to expect;
- Incorporating flexibility. This is particularly important when caring for a baby who will have their own schedule (or lack thereof). This will likely mean adjusting the plan along the way to account for needs of the family and the needs of the business;
- Communicating with staff and colleagues and delegating responsibilities whenever possible. This can include how/when to reach out, who else to call with questions, and which responsibilities will be completed by someone else. This can also include hiring relief/locum veterinarians to cover clinical responsibilities;
- Getting support from family and paid help. This can include having someone provide childcare for a few hours while administrative tasks are completed or while a few patients are seen, and delegating meal preparation and other household tasks;
- Setting realistic expectations and avoiding self-judgment. Not everything will go according to plan. Do not expect the time surrounding parental leave to be a period of significant business growth (although it may still be);

- Planning to sleep. Although it may be fleeting and hard to come by for both parents and business owners, sleep is essential;[23,24]
- Planning transition time to gradually reduce work responsibilities before giving birth and to gradually resume them when returning to work;
- Seeking support and guidance from others who have had similar experiences or are in similar life stages. This can help to prevent isolation, which can have negative effects on mental and physical health.[25]

Taboada counsels, "You are the biggest asset in your business. Investing in your well-being during this time will yield long-term benefits for your individual health as well as the health of your business."[25]

PARENTAL LEAVE FOR STUDENTS, INTERNS, AND RESIDENTS

Students who are expecting a child should plan with their university or program administration as early as possible to determine their options for taking leave after the birth of their child. Some students may elect to take a leave of absence during their pregnancy and rejoin their program after having their child. Others may choose to continue in their studies for as long as possible, with accommodations as needed, and plan to have their leave start closer to the time when their child is born. In some cases, the birth of a child may coincide with a planned break, such as a semester or summer break. In others, students may need to coordinate carefully with their instructors and school administration to determine how they will complete any coursework they miss during leave and/or whether accommodations can be made to allow them to bring their child to classes or study from home. If a pregnant student has taken a temporary leave of absence during their pregnancy, their maternity leave will likely be factored into this time. For non-birthing students, they will need to discuss whether taking a leave of absence, rearranging rotations, or other options are available to allow them to support their partner and bond with their child.

Veterinary interns and residents who become pregnant during their training can feel a lot of pressure to take the shortest leave possible to avoid overburdening the other members of their cohort. This is a concern among residents in human medical training as well. The residents in one study reported fearing the stigma associated with taking leave and worried that their knowledge, skills, and career advancement would suffer if they took more than a minimal amount of parental leave.[26] A survey of veterinary students, interns, and residents found that, as of 2019, only 13 of 26 US veterinary schools had any recommendation written in their handbooks regarding maternity leave. No veterinary schools had any written recommendations regarding paternity leave or leave associated with adoption. Only six schools reported having a specific written policy regarding parental leave for "house officers" (interns and residents). Only a few schools responded that they offered 6–12 weeks of leave or that interns and residents would be eligible for leave like any other employee. Other schools responded that leave would be arranged on a "case by case" basis, that this kind of thing "rarely happens," and that fathers usually take "less than a week" when it comes to paternity leave. One school reported that it would be "difficult to accommodate significant periods of time off."[27]

As training programs for medical students and new physicians are more commonly instituting family friendly practices, there is an opportunity for veterinary training programs to do the same. These programs may include formal, written parental leave policies, better lactation support, options to bring young children to lectures in a soundproof room when daycare is not available, and mentorship programs that help new parents navigate their dual roles.[27] In 2021, the American Board of Medical Specialties (ABMS) instated a policy requiring that training programs of at least two years' duration must allow for at least one period of at least six weeks of parental, caregiver, or medical leave "without exhausting all other allowed time away from training and without extending training."[28] A study of parental leave in medical trainees found that taking parental leave had no negative effects on trainees' knowledge, skills, or career progression when compared with trainees who took no leave during their training. It also suggests that the effects of the absence of a trainee on leave could be mitigated by paying other faculty members to cover the parent trainee's on-call shifts and/or asking for paid volunteers among the other trainees. This could lessen the burden and perceived unfairness of having to pick up extra shifts to cover for a colleague who is taking parental leave.[26]

BOX 9.4 Practice profile: Planning pregnancy and parental leave in vet school and as a business owner

Dani McVety, DVM

AS TOLD TO EMILY SINGLER

Dani McVety is a mom of four, a veterinarian, and the CEO and founder of Lap of Love Veterinary Hospice and In-Home Euthanasia. She knew that she wanted to have kids, but she didn't want to wait until after vet school to grow her family. Aware of the risks and challenges of navigating pregnancy during clinical rotations, she and her husband planned to try to conceive during her third year of school.

Dani remembers being the "first and only person to have a human child" in her vet school class. She didn't know anyone else who had done it, but she felt it was right for her. Since the first half of her third year at the University of Florida was in clinics, she hoped to be in the early part of her pregnancy then and be back in the classroom for the second half of the year when she was "big and uncomfortable." She explains that they had it "planned down to the day."

Luckily, their plan worked, and she got pregnant right away. That first half of third year in clinics, she was plagued with the typical exhaustion of the first trimester. She recalls having to go into a back room and put her head down and sleep. She carried around a water bottle to stay hydrated, but she didn't want others to have to make accommodations for her. She made sure to schedule her radiology and anesthesia rotations before getting pregnant. Looking back on it, she remembers having an easy pregnancy. Between the support of the vet school administration and the interest and attention from her classmates, Dani "felt like a goddess" during her pregnancy.

Her baby was born in June, and she was able to take the standard summer break from school for a three-month maternity leave. When she participated in her first externships when her son was three months old, her parents watched her baby. During the classroom portion of her fourth year, she had a predictable schedule and was able to pump regularly. She continued pumping through her first two rotations, but then felt comfortable stopping since her son was eight or nine months old by that time.

Dani recalls having to be very intentional with her time during this phase of her life. Since she relied on a babysitter for childcare, she had to fit everything she needed to do into the time in which the sitter was available. She had to become very regimented with her time and her studying. All-nighters were not an option. She failed the NAVLE the first time she took it. But she persevered.

With her subsequent pregnancies, she continued to be very busy, but had a little more control of her schedule. When her second child was born, she started seeing appointments with Lap of Love one week after giving birth, with her husband waiting in the car with their baby. After her third child was born, she started bringing the baby with her and asking clients if it was okay to bring the baby in for the appointment. Now, since her responsibilities are all managerial, she feels more flexibility to take a month just to develop a "flow" after she has a baby and then gradually increase her work responsibilities over the next few months.

At 3 PM every day, however, she steps away so that she can make it to her "next job": picking up the kids from school, talking to them about their day, and connecting with them. This is something she prioritizes every day, and she describes it as her "balance."

RETURN TO WORK OR GRADUAL REENTRY PROGRAMS

Although this terminology may not be well-recognized or commonly used in veterinary medicine, many other fields have adopted the practice of specifically and intentionally supporting parents as they return to work after leave. This is another benefit that employers can offer that can help improve employee retention and well-being. This type of benefit is particularly important for parents who may be experiencing sleep deprivation, mental health challenges, and general overwhelm as they adjust to caring for their new child while also trying to be successful in the workplace. Although the terms "employer" and "employee" are used, the concepts are equally applicable to students returning to their field of study. Vivvi states, "A truly inclusive organization should give new parents the tools they need to return with confidence."[29] Just as many companies will need to tailor their parental leave policies to the needs of their employees and their business, every return-to-work program may look different as well. Some possible components include:

- Just as an intentional plan is put in place to transfer and/or pause any projects and responsibilities under the purview of the parent going on leave is important, so is an intentional "on-ramping" plan to help the newly returned employee gradually get back into the swing of things and gradually start taking on their tasks and responsibilities again. This can also allow the employee to be brought up to date on any staff, policy, or schedule changes that have occurred in their absence;

- A gradual return-to-work plan can allow employees to return to work at less than their full-time hours at first to help ease their stress levels with regard to separation from their child, lack of sleep, adjusting to managing childcare, and focusing on work responsibilities again. In some situations, employees may be able to return remotely, completing tasks such as client communications, refill requests, and telehealth visits;
- Encouraging empathy toward the stresses of caring for a young child and coming back to work on the part of all colleagues is very important. One way to foster this is to ask questions that allow new parents to share their experiences and perspective so that their coworkers can have a better understanding of their needs. Being a new working parent is stressful, even if the employee is not experiencing postpartum depression or anxiety. It should be made clear that any micro-aggressions, including jokes about the employee having "been on vacation," "having abandoned" the workplace, or "getting special treatment" due to their needs as a parent are inappropriate and will not be tolerated;
- Some workplaces offer employer-sponsored childcare benefits. This can be in the form of an on-site daycare at the workplace, a stipend that can be used to help finance childcare, negotiating a corporate discount with certain childcare providers, offering a tax-advantaged dependent-care flexible savings account, and/or offering access to programs that provide emergency back-up childcare;
- Peer mentoring programs (pairing newly returned parents with other parents in the workplace) can help ease concerns about how having children affected their career. This can help new parents learn from others who have been through similar experiences for the benefit of moral support. These experienced mentors can also help guide newer parents to feel successful and continue to grow professionally as well.[29]

TAKE-HOME POINTS

- Parental leave is essential for healing, bonding, establishing childcare routines, and supporting parents' mental health;
- Providing paid parental leave can significantly improve employee well-being and retention and make parents much more likely to take the time they need to heal and bond;
- The whole family and the workplace will benefit when both parents (in a two-parent household) have equal access to paid leave and feel empowered and encouraged to take it;
- Paid leave can be funded in different ways: at the country or state level, by the individual company, and through short-term disability insurance. Unfortunately, benefits differ significantly between countries, states, companies, and even within companies;
- Planning leave early will allow time to explore all available options and save up money for self-funded leave if necessary;

- In areas where no national or state paid leave program exists, small businesses can still find creative ways to offer paid leave to employees and to business owners;
- There is room for improvement in parental leave policies offered in veterinary schools and training programs;
- Return to work and gradual reentry programs can help reduce stress and anxiety around coming back to work after leave and can improve a new parent's appreciation for and success in the workplace.

REFERENCES

1. Mercer. Want to improve gender equality at work? Help men take parental leave. October 30, 2018. Accessed November 6, 2022. https://www.mercer.com/our-thinking/want-to-improve-gender-equality-at-work-help-men-take-parental-leave.html#:~:text=Maothers%20are%20also%20half%20as,month%20of%20leave%20men%20take

2. Colantuoni F, Rajbhandari S, Tolub G, Diome-Deer W, Moore K. A fresh look at paternity leave: Why the benefits extend beyond the personal. March 5, 2021. Accessed November 6, 2022. https://www.mckinsey.com/capabilities/people-and-organizational-performance/our-insights/a-fresh-look-at-paternity-leave-why-the-benefits-extend-beyond-the-personal

3. National Partnership for Women and Families. Paid leave is essential for healthy moms and babies. May 2021. Accessed November 6, 2022. https://www.national-partnership.org/our-work/health/moms-and-babies/paid-leave-is-essential-for.html

4. Wang W. Parents' time with kids more rewarding than paid work—And more exhausting. October 8, 2013. Accessed November 30, 2022. https://www.pewresearch.org/social-trends/2013/10/08/parents-time-with-kids-more-rewarding-than-paid-work-and-more-exhausting/

5. Andersen SH. Paternity leave and the motherhood penalty: New causal evidence. *J Marriage Family*. 2018;80(5):1125–1143.

6. Organisation for Economic Cooperation and Development. Parental leave: Where are the fathers? March 2016. Accessed November 16, 2022. https://www.oecd.org/policy-briefs/parental-leave-where-are-the-fathers.pdf

7. Bipartisan Policy Center. Paid family leave across OECD countries. March 1, 2022. Accessed November 16, 2022. https://bipartisanpolicy.org/explainer/paid-family-leave-across-oecd-countries/

8. World Population Review. Maternity leave by country 2022. Accessed November 9, 2022. https://worldpopulationreview.com/country-rankings/maternity-leave-by-country

9. Michelson J. How small companies can offer great paid-leave programs. January 7, 2021. Accessed November 6, 2022. https://hbr.org/2021/01/how-small-companies-can-offer-great-paid-leave-programs

10. Parentaly. How small companies can offer paid parental leave. March 19, 2022. Accessed November 6, 2022. https://www.parentaly.com/blog/small-companies-paid-parental-leave

11. infoFinland.fi. Family leave. Accessed November 23, 2022. https://www.infofinland .fi/en/work-and-enterprise/employees-rights-and-obligations/family-leave

12. National Conference of State Legislatures. State family and medical leave laws. Accessed November 9, 2022. https://www.ncsl.org/research/labor-and-employ-ment/state-family-and-medical-leave-laws.aspx

13. U.S. Department of Labor. Family and Medical Leave Act. Accessed November 16, 2022. https://www.dol.gov/agencies/whd/fmla

14. Fairygodboss. Paid maternity leave: 180 companies who offer the best paid leave. February 13, 2022. Accessed November 16, 2022. https://fairygodboss.com/articles /paid-maternity-leave-companies-who-offer-the-most-paid-leave#

15. Department of Commerce. Paid parental leave for federal employees. Accessed November 16, 2022. https://www.commerce.gov/hr/paid-parental-leave-federal-employees

16. Today's Veterinary Business. Pathway provides 8 weeks of paid parental leave. January 17, 2020. Accessed November 6, 2022. https://todaysveterinarybusiness .com/pathway-provides-8-weeks-of-paid-parental-leave/

17. Peralta P. The veterinary clinic is covering their employees' healthcare in full. September 3, 2021. Accessed November 6, 2022. https://www.benefitnews.com/ news/bond-vet-is-covering-their-employees-healthcare-costs-in-full

18. Vet Partners. Paid parental leave. Accessed November 6, 2022. https://vet.partners/employee-benefits/paid-parental-leave/?utm_source=VetPartnersWebsite&utm_medium=EmployeeBenifitsPage?&utm_campaign=VPCareersPages

19. Cision PR Web. CareVet launches best-in-class veterinarian benefits, including paid parental leave, adoption assistance, student loan support, paid luxury vacations and more. November 8, 2021. Accessed November 6, 2022. https://www.prweb.com /releases/carevet_launches_best_in_class_veterinarian_benefits_including_paid _parental_leave_adoption_assistance_student_loan_support_paid_luxury_vaca-tions_and_more/prweb18316539.htm

20. Vet Practice. Supporting working parents. June 15, 2020. Accessed November 6, 2022. https://www.vetpracticemag.com.au/supporting-working-parents/

21. Reaume A. Will short-term disability insurance cover maternity leave? June 20, 2022. Accessed November 16, 2022. https://www.northwesternmutual.com/life -and-money/will-short-term-disability-pregnancy-maternity-leave/

22. Beacom, A and S Campbell. *The Parental Leave Playbook*. Hoboken (NJ): John Wiley & Sons; 2021.

23. Huang G. How to manage a short maternity leave if you're self-employed. Accessed November 28, 2022. https://www.forbes.com/sites/georgenehuang/2018/01/02/how -to-manage-a-short-maternity-leave-if-youre-self-employed/?sh=119b0c6320ba

24. Lord M. A guide to maternity leave for entrepreneurs. October 14, 2021. Accessed November 28, 2022. https://www.entrepreneur.com/leadership/a-guide-to-mater-nity-leave-for-entrepreneurs/237957

25. Taboada A. *The Expecting Entrepreneur: A Guide to Parental Leave Planning for Self Employed Business Owners*. Berkeley: Arianna Taboada; 2021.

26. Huh DD, Wang J, Fliotsos MJ, et al. Association between parental leave and ophthalmology resident physician performance. *JAMA Ophthalmol.* 2022;140(11):1066–1075.

27. Molter B, Wayne A, Mueller MK, Gibeley M, Rosenbaum MH. Current policies and support services for pregnant and parenting veterinary medical students and house officers at United States veterinary medical training institutions. *J Vet Med Educ.* 2019;46(2):145–152.

28. American Board of Medical Specialties. American board of medical specialties policy on parental, caregiver and medical leave during training. July 1, 2021. Accessed November 28, 2022. https://www.abms.org/policies/parental-leave/

29. Vivvi. Beyond maternity: How employers can support mothers returning to work. April 18, 2021. Accessed November 28, 2022. https://vivvi.com/blog/articles/supporting-mothers-returning-to-work

"Mom, what did you do all day?"

The fourth trimester and parental leave

When I went on maternity leave after the birth of my first child, my world was turned upside down in so many ways. In addition to healing and learning how to care for my newborn child, I had to cope with the complete loss of structure to my day. Whereas I would previously follow a schedule of waking up, eating, going to work, seeing appointments, and coming home, my days now seemed to have no beginning or end. I had to create new routines, and then be prepared to totally change them when my baby decided she wasn't on board. I had to accept help. I had to give up on perfectionism. I had to learn about breastfeeding and pumping and bottles and how to get my baby to sleep when I wasn't holding her. And I had to keep her safe, fed, and isolated from infectious disease, and deal with all the new (and often overwhelming) emotions and anxieties that come with having a new baby. And then, when I was just starting to get the hang of it, I had to figure out how to do all of that and activate my brain to practice veterinary medicine again. I am incredibly grateful to have not had any serious complications with the births and health of any of my children. I am painfully aware that this is not every family's experience.

DOI: 10.1201/9781003406907-10

NAVIGATING THE FOURTH TRIMESTER

While pregnancy encompasses three trimesters, the fourth trimester is defined as the first 12 weeks after a baby is born.[1] These first few months after birth can be filled with sleepless nights; pain and other physical challenges associated with recovery from childbirth; hormonal changes; baby blues; stress and worry over the baby's health, development, and feeding; postpartum depression and anxiety; and, for some, concerns about having to return to work.

Childbirth and the peripartum period can be challenging in even the best scenarios. In some cases, things do not go as planned, and either the mother's and/or the baby's life or health can be threatened. Some babies are born prematurely with health challenges that require additional treatment, extended hospital stays, and time in a neonatal intensive care unit (NICU). Some parents will navigate the postpartum period without a baby (e.g., placement of the baby for adoption, stillbirth, or death at or shortly after birth). These parents still need all the physical healing, often experience all the hormonal changes as if they were still caring for a baby, and then on top of that, they need to grieve the loss of their baby or babies.

Just as every family and every baby is different, every fourth trimester experience will be unique. This can include the community and other resources available. Some parents will have copious family and/or community support to help both parents and baby rest, recover, and bond. Others may have access to a postpartum doula, a visiting midwife, a night nurse, a health visitor, or other professional help. Still others may, by choice or necessity, have very limited outside support. New parents are encouraged to take advantage of any resources available to them and to remember that they were never meant to weather the challenges of postpartum life alone.

Because it is outside the scope of this book to cover parenting, childcare, and the myriad health considerations associated with the postpartum period in detail, the reader is encouraged to review this material from other reputable sources (see the appendix at the end of the book for some recommendations) and to seek advice from their own healthcare providers. This chapter will include a small selection of some of the challenges parents and infants face in the fourth trimester period, methods to navigate them, and a discussion about boundaries and preparing to go back to work. Mental health will be covered separately in the next chapter.

PARENTAL LEAVE

For working parents, their parental leave may coincide exactly with the fourth trimester, or it may extend to a period of time before and/or after the first 12 weeks have ended. Especially in situations where time and/or pay for maternity leave are limited, many expectant parents may try to use as little as possible of their leave before the birth of their baby to maximize the amount of time available after the baby is born. However, in some countries and with some employers, it may be common or even expected for pregnant individuals to begin their leave before their expected due date, even when no pregnancy complications are present. In other cases, the health of the baby and/or mother may warrant an earlier start to maternity leave, or a pregnant individual may decide to begin leave early for other reasons. Whatever the reason, this can be a time to prepare for childbirth, and for a new baby at home, to avoid overexertion and exposure to contagious disease, and (hopefully) to rest.

Parental leave after the birth of a baby is important for many reasons. Mothers in the postpartum period need at least six weeks to completely heal from childbirth, no matter how the infant was delivered, even if they had no complications. Spending sufficient time at home before returning to work can allow parents—whether they delivered a baby or not—to establish a new routine, bond with their child, and find ways to incorporate self-care into their lives before having to resume their workplace responsibilities. Parental leave is also very important for babies who are vulnerable and need round-the-clock care. When a parent is home, their newborn baby can remain at home and not be as exposed to contagious disease as they would be in an alternative childcare setting. They can consistently establish feeding and sleeping habits. They can bond with their parents and form a strong attachment.

Despite all of these benefits, there is no one-size-fits-all solution for the duration of parental leave. Some individuals will feel so financially constrained that taking even six weeks of leave is not possible. Others, especially practice owners, will feel a need to get back to work in some capacity to resume their managerial responsibilities as soon as possible. Others will decide to return to work early to support their own mental health.

BOX 10.1   Workplace culture and maternity leave

Jen (last name withheld), DVM

As told to Emily Singler

Dr. Jen is a small animal relief veterinarian with two children. She tells of two very different experiences on maternity leave that were affected by the level of support and the culture of her two different work environments at the time of each pregnancy. Also very important in her experience was the support of her husband and her husband's employer in permitting him to have a paid paternity leave with the birth of each child.

Jen and her husband battled years of infertility, underwent multiple treatments, and had multiple early pregnancy losses. She describes working in a practice while undergoing fertility treatment where she was made to feel guilty about having to miss work for doctor's appointments, and at one point was accused of having lied about having a miscarriage in order to miss work.

While working in another practice, she found out she was pregnant, and after a scare about another loss, she learned that her pregnancy was viable. Initially she believed she had a verbal agreement with her employer that she would be able to have a certain amount of maternity leave and that she would be able to use paid time off for a portion of it. She now feels like she assumed too much, as she was met with a very different attitude one month before her leave was to start. As a result, she was only able to take a six week leave with only one week paid. She felt that the stress of these changes shortly before the birth of her first child may have contributed to her going into labor three weeks before her due date.

While on leave, she felt a sense of relief to be home and not working in the practice where she experienced so much stress. She was very excited to finally be a mom, so she was able to shut off any thoughts about work. But she was overwhelmed with the responsibilities of caring for a newborn mainly due to sleep deprivation. After six weeks, not enough time in her words, she returned to work. On her return to work she was denied any scheduled pumping breaks and criticized for not working more during her annual review.

During her second pregnancy a few years later, Jen was doing relief at a different practice. She remembers being invited by her employers to sit down and discuss her needs and wishes at two different times during her pregnancy. She was encouraged to do what made her feel comfortable in terms of doing surgery while pregnant, scheduling patients and planning for leave. She was much less stressed during the duration of her pregnancy, and she feels this helped her carry her baby to term. As excited as she had been when her first child was born, she describes being even happier on her second maternity leave. She was able to use paid time off accumulated from the previous year and also borrow some from the year ahead. As a result, she had five weeks of paid time off and a total of nine weeks' maternity leave. This felt perfect to her, and she was ready to go back to work two days a week after nine weeks of healing and bonding. When she returned to work, she was very relieved to see that her pumping breaks were treated as "sacred," the total opposite from her previous return to work.

Jen describes having a good network of friends and family during both maternity leaves. Her husband was very involved and made efforts to share the childcare load. This included middle of the night feedings once a bottle could be given. He was also fortunate to be able to take some paid paternity leave time for the birth of each child. His state job offered up to 12 weeks of paternity leave to be taken all at once or in parts during the 12 months after the birth of a child, and he was allowed to amass his sick days to use for this purpose. Since he had been at his job for 15 years, he was able to have enough time to take off 2 weeks when their first child was born and then another 2 weeks to stay with the baby when Jen first went back to work. For their second child, he took a month of paid leave when their baby was born and another two weeks when Jen went back to work. Jen remembers what a huge difference this made for her well-being, as it allowed her to feel less overwhelmed, get more sleep, and have a more equal distribution of chores like dish washing and grocery shopping.

MATERNAL HEALTH

POSTPARTUM/POSTNATAL HEALTH CARE

The American College of Obstetricians and Gynecologists (ACOG) identifies some of the postpartum challenges in the fourth trimester, including "lack of sleep, fatigue, pain, breastfeeding difficulties, stress, new onset or exacerbation of mental health disorders, lack of sexual desire, and urinary incontinence." ACOG also indicates that postpartum medical care in the US is often disjointed, lacking in communication, and almost entirely the responsibility of the healing new mother to coordinate. As a result, only about 40% of people in the US keep their postpartum appointment. Even those who do keep their appointment may be missing important medical care. Although a single postpartum appointment at 6 weeks post-birth has been the norm, the recommendation from ACOG is to have multiple visits, starting within 1 to 3 weeks of birth and another within 12 weeks of giving birth. More frequent visits may be needed when health problems are present. These visits can sometimes be by phone or video, but at least one visit should be in person to allow for a physical examination. Postpartum visits are vital to ensuring the mother is recovering well physically, that their mental health is addressed, and that their coping strategies and support systems for weathering the rigors of postpartum life are healthy and intact.[2]

In the UK, a health visitor is "allocated to every family with a new baby to give support to mother and baby and protect child and family health and wellbeing." This support continues until the child reaches five years of age and starts school. Areas of support can include feeding, vaccines, safety, postnatal medical care, family planning, mental health concerns, safe sleep, and other topics as needed by the family. In addition to the health visitor, a midwife may continue to visit families of newborns at home. The goals of postnatal care, according to the National Institute for Health and Care Excellence (NICE), are that:

- Women feel they are listened to and their postnatal care is driven by what matters to them
- Women get better information and support to care for themselves and their baby, including how to stay safe if they share a bed with their baby
- Women always get the necessary information and support with feeding their baby
- Postnatal visits are planned and spaced out better to avoid gaps in support.[3]

BOX 10.2 Practice Profile: The fourth trimester

Alyssa Allen, DVM

The fourth trimester is a wonderful and challenging time of your life. For me, it started with more pain than I expected, considering a very easy delivery. My husband ended up changing not only all of my baby's diapers but also a bunch of mine the first few days. When I look back, I remember the pain but also the love. The love of finally understanding how indescribable the love for your child is. The love of seeing your partner become a caring parent. I would sit and do skin to skin with tears in my eyes knowing my son would never be that small again and sad at how quickly my leave was going.

The first month I remember so much fluid, which I was not prepared for. Between lochia, leaking breast milk, and night sweats, it was a lot. My milk luckily came in strong… a little too strong as I got mastitis two weeks in. A Hakka on one side helped build a stash while also not soaking burp cloths and didn't lead to oversupply. I had to plan showers for just before nursing, because the warm water caused a letdown and I wanted the leaks to go into my baby instead of the towel.

Soon, the sleep deprivation kicks in. It was a hard few weeks, but you will get through it. We found trading off worked; my husband would change each diaper while I got situated in nursing position, and he'd bring my son right to me in whichever way he was going to nurse. It made us laugh, and then he'd go back to sleep. During the day, he'd take the first morning awake shift so I could sleep "in" between the feedings. The morning light this year brought back memories of that time, and when you are in the thick of it, it will be hard. But know that you will likely remember it fondly.

The next two months were some of my favorite months ever. I had the privilege of having my husband home for 10 out of 13 weeks, meaning we spent months 2 and 3 together. We like to get out of the house, so we became experts at being on the go. The last 2 weeks we took a road trip through the Midwest, stopping every 1–3 days in a new place, traveling 3000 miles to see as many friends and family as possible. We were a bit crazy because our two dogs came as well as our Snoo bassinet, jogging stroller, and many other things we probably didn't need. I found that getting out and doing things that were important to us while finding a way for a baby to fit in made our first year more enjoyable. A baby doesn't have to stop your fun or spontaneity, but you may have to find ways to make slight changes. It is impossible to be completely prepared, but you adjust and become the best mom for your child.

MATERNAL MORTALITY

Although the Centers for Disease Control and Prevention (CDC) report that "about 700 women die each year in the US as a result of pregnancy or delivery complications," almost half of those maternal deaths happen after the baby is born.[4,5] Columbia University explains that some postpartum deaths happen as late as one year after delivery. Black people have three to four times the risk of dying of a pregnancy-related cause than white people, and the maternal mortality rate for non-Hispanic, American Indian, or Alaska Native persons is twice as high as it is for white individuals.[1,4] NICE confirms that maternal mortality is four times higher for Black people than white people in the UK as well, and three times higher for "mixed ethnicity women."[2] The CDC identifies possible causes for this devastating disparity, which may include "access to care, quality of care, prevalence of chronic diseases, structural racism, and implicit biases." These factors not only influence the types of care that are available, but also the ways in which people's needs, concerns, and even vital signs are listened to and interpreted.

The most common causes for pregnancy-related maternal deaths reported by the CDC include:

- Infection or sepsis;
- Cardiomyopathy and other cardiovascular conditions;
- Hemorrhage;
- Thrombotic pulmonary or other embolism (blood clots);
- Cerebrovascular accidents (strokes);
- Hypertensive disorders of pregnancy (high blood pressure);
- Amniotic fluid embolism;
- Other non-cardiovascular conditions.[4]

Columbia University lists warning signs of potentially serious health problems in post-partum mothers that should not be ignored because they may be a sign of a life-threatening complication. They include:

- Fever or chills;
- Heavy, brisk, bright-red bleeding (saturating a pad every hour for several hours);
- Dizziness;
- Shortness of breath or chest pain;
- Severe headache;
- Swelling of the legs and feet;
- Deep sadness; thoughts of hopelessness; thoughts about death, suicide, or harming oneself or the baby.

Mothers with diabetes, high blood pressure, obesity, or other chronic health conditions are at higher risk of the development of a health problem and should take any signs that they notice very seriously.[1] Even for problems that may not be life-threatening, seeking prompt medical care is very important to help promote good quality of life.

In a study by the Commonwealth Fund, maternal mortality ratios (the number of deaths of people during pregnancy or within 42 days after pregnancy per 100,000 live births) was compared among 10 developed countries.[6] New Zealand and Norway had the lowest maternal mortality rates at 1.7 and 1.8 deaths per 100,000 live births, whereas the US had the highest rate at 17.4 deaths per 100,000 births. The authors contribute the disparity among these countries to their differences in maternal and postpartum care. The US is the only country among the ten that does not routinely utilize home visits to postpartum mothers.

In all of the other countries in the study, a midwife or nurse typically visits the mother at home within 1 week of delivery, sometimes within 24 hours. These visits are also covered by national health insurance in all countries. The US also has the lowest ratio of midwives available per 1000 live births (tied with Canada). This is likely the case because most private insurance does not cover care by a midwife outside of the hospital setting and many practice acts may limit the scope of care midwives can provide. However, the study authors report that "countries with the lowest intervention rates, best outcomes, and lowest costs have integrated midwifery-led care into their health systems."[6] Postpartum parents who have access to midwives and other healthcare professionals for ongoing care and monitoring are encouraged to take advantage of it for their own health and well-being.

Steps to Maintain Good Postpartum Health

It can sometimes be hard for parents who have recently given birth to remember that, in addition to caring for their new baby, they need to take care of themselves. Stanford

Medicine stresses that the most important need of postpartum birthing parents is rest. Although it may seem to be the most elusive of goals, rest is essential to be able to manage the physical and mental stresses placed on the body during this time. It will undoubtedly not happen in eight-hour increments, but it is important for parents to aim to get as much rest as possible after giving birth. There are multiple ways to increase the likelihood that this happens. These include:

- Not performing any tasks at home that are unrelated to caring for the baby or themselves, such as cooking, cleaning, shopping, driving to activities, cleaning bottles and pump parts, or any other task that can be delegated to others in the first few weeks;
- Sharing infant care responsibilities as much as possible with a partner and any other trusted helpers;
- Trying to follow the cliché "sleep when the baby sleeps." It doesn't always work, but even short naps can be restorative;
- Having the baby's crib or bassinet near the bed for ease of access at night;
- Not feeling the need to entertain any guests who come over. They can hold the baby or do the dishes while mom takes a shower or a nap.

Nutrition is another key part of self-care for postpartum parents. It can be hard to prepare a meal or even feel like eating when sleep and day-to-day schedules have been completely thrown off. Some people feel an intense pressure to lose weight quickly after giving birth, and this may drive them to cut back on their eating. But good nutrition is essential not only for breastfeeding (where applicable) but also for recovery from childbirth and for managing the stresses of caring for a baby practically 24 hours a day. Stanford Medicine recommends planning "simple, healthy meals" that include grains, vegetables, fruits, dairy, and protein. Drinking plenty of water is also very important, especially for those who are breastfeeding.[7]

Spending short periods of time outside every day and engaging in light exercise once a postpartum mother has been cleared to do so can help support both physical and mental health. In addition, postpartum parents should stay in tune with their bodies and reach out to their healthcare provider with any questions or concerns, especially if they are at high risk or if they notice any of the health problems listed previously.

INFANT HEALTH

NICU STAYS

There are many reasons why a baby might need to spend time in the NICU. According to Stanford Medicine, these can be for maternal reasons (such as high blood pressure and premature rupture of membranes), delivery complications (such as a breech birth or fetal distress during delivery), and newborn complications (such as birth defects, premature birth, low birth weight, respiratory distress, and low blood sugar). It is very common that babies who are born prematurely and/or who are multiples must spend some time in the NICU. A NICU stay can help support a newborn baby while their lungs and other organs continue to mature. It can be a place where extra nutritional support can be given to babies who cannot drink milk or formula on their own or who cannot maintain adequate blood sugar levels. It can also provide life-saving and life-sustaining support to babies with birth defects and other conditions while they receive treatment. Babies who spend time in the NICU will have a team of highly qualified professionals caring for them around the clock. These typically include neonatologists, pediatricians, nurses, respiratory therapists, physical therapists, occupational therapists, speech therapists, dietitians, lactation consultants, pharmacists, social workers, and more.[8]

Having an infant (or infants) in the NICU can be very scary and stressful for parents. They will naturally worry about the health and safety of their child(ren). It can be terrifying for a parent to see their tiny baby hooked up to so many tubes and wires. Parents may not be able to hold or interact with their baby as much, or they may not be able to hold their baby at all. This can be extremely painful for parents who want nothing more than to nurture and protect their child. For families of multiples, it can feel like an impossible challenge to determine how to split their time and attention between more than one baby.

Additional heartbreak can ensue when a mother is discharged from the hospital, but their baby is not. Parents may no longer have a place to stay overnight in the hospital, and they may be forced to travel back and forth frequently to sleep and/or care for other children and pets and then return to the NICU to spend time with their baby. In addition to sleep deprivation and hormonal changes, postpartum mothers may spend much of their waking hours trying to express breastmilk for their baby. It can be very hard postpartum for a mother to rest and work on healing when their child is in the NICU. Some may have the added pressure of having to return to work, particularly if they had to start their maternity leave before their baby was born.

The March of Dimes reports that there are ways for parents to take care of themselves while their baby is in the NICU. These include:

- Creating routines that include eating, hydrating, taking a shower, and sleeping every day. Parents can decide when in this schedule they will spend time with their baby;
- Taking breaks from the NICU for self-care and spending time with other children and their partner;
- Seeking others to connect with who have had similar experiences with babies in the NICU and find support and advice;
- Seeking support from a mental health counselor or a religious or spiritual leader;
- Asking for help from friends and family. This can include delegating meal prep, grocery shopping, and other household tasks; help with other children; and even taking shifts in the NICU to allow a parent to take a break;
- Communicating with employers if the parent has returned to work. Parents may be able to request a schedule change, reduced hours, or other accommodations if needed.[9]

The March of Dimes also recommends trying to be involved in the baby's care as much as possible when they are in the NICU. This may include holding, feeding, changing, bathing, weighing, and other care as deemed appropriate. For parents who have been discharged from the hospital, some may be able to stay overnight in the hospital in rooms designated for NICU parents, or there may be free or low-cost lodging close to the hospital for NICU parents to stay.[10]

BOX 10.3 Practice profile: Our NICU story

Jessica Smith, DVM

My precious Autumn Lynn, you were born at 34 weeks 6 days. You were 17 3/4 inches long and weighed exactly 5 pounds. I got to spend what felt like the fastest ten minutes of my life holding you before we were separated and you were whisked away to the NICU, surrounded by strangers and without me by your side.

I was at work seeing appointments and had been feeling lightheaded and dizzy at times throughout the morning. I went to my routine 34-week appointment that same day, and my blood pressure was high. My doctor recommended I keep monitoring my blood pressure at home, and it stayed high. We were told to go see a specialist, and from there, admitted to the hospital. I was told that I had preeclampsia, and that I needed to get induced. After a difficult induction, your birth was a blur.

They said you had to go to the NICU because you were a preemie. When they handed you to me, you were already connected to monitors and a CPAP but you were the most beautiful baby. You stared up at me and that is the moment you became my whole world. After just ten minutes, they said they had to take you to the NICU. Your dad didn't even get to hold you before they took you from me and wheeled your isolet out of the room. I asked to go with you but they said I had to wait until I was brought up to a different hospital room.

The feelings I had in that moment were indescribable. I felt a sense of loss, panic, and anger that I was separated from my newborn. As soon as they got me transferred to a new room, the nurse brought me to see you. The NICU felt like another world. You were so small in your bed. They had already started an IV but you were no longer on CPAP. I finally got to hold you again and I told you I would fight to get you released. I asked the nurse how long you'd be there but she told me they didn't know yet. I held you as long as I could before the nurse said I needed to go back to my room. I gently laid you in your bed and cried the whole way back to my room because I felt like the worst mother for leaving you alone in a strange, scary place that we both didn't understand.

I came to visit you as often as I could while I was hospitalized but two days after you were born, they said I was going to be discharged. I was both relieved and extremely anxious because I knew I wouldn't get to take you with me. I remember going home and walking upstairs to your nursery and collapsing on the floor crying because you weren't there and your dad drove me back to the hospital so I could sit in the NICU with you again. The two weeks you were in the NICU felt like an eternity. It was the worst emotional rollercoaster experience I've ever had. We would wake up early, let the dogs out, and then go to the hospital where we would sit in the NICU all day and take turns holding you.

Every day I would ask the nurses how you were doing and when you could go home. They would always say it depended on you. Some days they would be very encouraging and say you were exceeding their expectations and other days they would say it could be a while before you were ready. The biggest reason you were in the NICU was because you wanted to sleep through your feedings so they had to place a feeding tube. I quickly learned that the amount you would eat from the bottle depended a lot on your nurse. Some were kind and patient with you and would give you as much time as you needed before feeding the rest through the tube but others would give up quickly and wouldn't let you try. On days where I didn't like your nurse, we would stay the whole day until shift change so that we were the ones doing your care times. I felt so protective of you and yet so helpless because I couldn't get you out of the NICU.

Finally, exactly two weeks after you were born, the doctor called us while we were on the way to the hospital and said you could be discharged. We were so excited! I sat impatiently through the doctor's discharge and the nutritionist's explanation of how to feed you. It felt surreal to put you in your car seat and walk out of the hospital with you. It was one of the happiest days of my life! I would never wish the NICU experience on any parent and I feel so sorry for the parents who have to wait even longer before their baby is released.

However, the NICU did teach me some things that I will always carry with me. It taught your dad and me how to care for a newborn. We were so excited to take you home, but we truly had no idea how to care for you so I will be forever grateful for the patient nurses who taught us. The NICU also showed me that I needed to fight for you and be your advocate. I was one of "those" parents who demanded to speak with your doctor every day and wanted an explanation of the plan for your care. Finally, the NICU showed me what an amazing support system we have. We had people from church bringing meals, others were stopping by to let the dogs out and feed them, and still others would just check in to see how you were doing as well as offer encouragement. We never could have made it through that time without everyone's help. I am so grateful to have you home, Autumn Lynn. I will never stop loving you and fighting for you and, most of all, I'll never stop being there for you as your mom.

INFANT CARE AT HOME

Newborn babies are still very vulnerable as their body continues to grow and develop after birth. According to the 4th Trimester Project, "Babies depend completely on their caregivers for comfort, care, love and nutrition."[11] Newborn and infant care involves feeding (breastfeeding, formula, and/or tube feeding, and later, solid foods), ensuring safe sleep, establishing connection and a bond, and protecting babies from infectious disease. Parents should consult regularly with their pediatrician or other healthcare provider to ensure their baby is healthy and to address any concerns. A small selection of important infant health topics will be reviewed here, but this is by no means an exhaustive list.

ESTABLISHING FEEDING

Feeding can be one of the most frustrating elements of infant care. Although term babies are usually born with the instinct to feed right way, there can be multiple reasons for them to have trouble with breastfeeding and/or bottle feeding. According to the US Department of Agriculture and the National Health Service, these challenges can include:

- Low milk supply;
- Tongue-tie and other anatomic changes that make it harder for babies to latch sufficiently;
- Gastroesophageal reflux;
- Cluster feeding;
- Pain with breastfeeding;
- Clogged milk ducts;
- Mastitis;
- Fungal infections;
- Exhaustion;
- Difficulty transitioning from the breast to the bottle;
- Negative effects to the mother's mental health caused by any complications or challenges associated with feeding.[12,13]

Some of these challenges may be temporary and may resolve over time. Many, however, will benefit from some type of intervention. The Lactation Network recommends that all postpartum mothers who choose to breastfeed consult with a lactation consultant in addition to their other healthcare providers with any concerns regarding feeding. Even mothers who are not experiencing obvious challenges can benefit from an evaluation of the baby's latch and to rule out any evidence of a tongue-tie or other anatomic changes. A lactation consultant can recommend steps to improve latch, the position in which the baby is held, pumping techniques, and more.[14]

Some babies will not breastfeed or will not continue to breastfeed for a variety of reasons, including poor milk supply, poor latch, preference for a bottle, difficulty with frequent pumping, the mother's mental health challenges associated with breastfeeding, and infant health problems precluding successful breastfeeding, among others. The American Academy of Pediatrics (AAP) advises that "when needed, infant formulas can provide excellent nutrition for your baby." AAP also recommends that all infants under the age of 12 months who are fed formula only or partially fed formula be given an FDA-approved iron-fortified formula, because some babies do not have sufficient iron reserves.[15]

INFECTIOUS DISEASE PREVENTION

Of particular concern for newborn babies and infants are infectious diseases that can be contracted from adults and/or other children. Although the adult or child carrying the disease may show little to no signs of illness, newborn babies are very vulnerable because of their immature immune systems. These diseases can include the "common cold" (which can be caused by one or more of 200 respiratory viruses, including rhinovirus and respiratory syncytial virus [RSV]), influenza, pertussis ("whooping cough"), and others.[16] The CDC report that between 2010 and 2020, as many as 20 babies under 3 months of age died from pertussis each year in the US.[17-19]

There are multiple steps that can be taken to reduce the risk of life-threatening infections in newborn babies. These include insisting that all individuals wash their hands before touching the baby. Hand sanitizer can be helpful if water and soap are not available. Any people showing signs of infectious disease should stay away from the baby. Avoiding crowded public places (as much as possible) for the first few months can also limit baby's exposure to contagious disease.[16]

The CDC also advises that "a baby's family members and caregivers should be up to date on their vaccinations to help form a circle of disease protection around the baby." This includes parents, siblings, grandparents, babysitters and nannies, and others. Recommended vaccines are the flu vaccine (during flu season) and the whooping cough vaccine (DTaP for children and Tdap for others). These recommendations are most critical for families of babies younger than six months of age, because they are too young to receive the flu vaccine and are at risk for serious complications if infected. Individuals in need of vaccination should receive them at least two weeks before meeting the baby to allow the vaccine to stimulate protective antibodies. It is recommended that most pregnant people be vaccinated during their pregnancy to provide their baby some initial immunity in utero.[20]

SUDDEN UNEXPECTED INFANT DEATH AND SAFE SLEEP

According to the CDC, sudden unexpected infant death (SUID) is a catch-all term that encompasses every sudden, unexpected death of an infant less than one year of age "in which the cause was not obvious before investigation." This affects approximately 3400 infants per year in the US. Sudden infant death syndrome (SIDS) is included in this category, along with accidental suffocation or strangulation in a sleep environment and other unknown causes. With more emphasis on safer sleep practices, SUID rates have fallen by more than 60% since the 1990s.[21]

To help reduce the risk of SIDS and accidental suffocation or strangulation, the National Institutes of Health created the "Safe to Sleep" campaign[22] based on guidelines from the AAP's Task Force on SIDS. The principles of safe sleep include:

- Placing babies to sleep on their backs for naps and at night;
- Using a firm, flat sleeping surface covered only with a fitted sheet;
- Feeding breastmilk for as long as possible, even if not exclusively;
- Having the baby in the same bedroom as the parents in their own sleep space for the first six months;
- Not placing any objects, toys, bumpers, or blankets in the crib;
- Offering the baby a pacifier for naps and at night (not attached to clothing or anything else);
- Not smoking or vaping in the home or around the baby;
- Not consuming drugs or alcohol while caring for the baby (especially if co-sleeping);
- Avoiding overheating and over-bundling the baby;
- Keeping the baby's head and face uncovered during sleep;
- Not letting the baby sleep or otherwise be alone in a vehicle;

- Seeking regular health care for the baby and following advice on prevention and treatment;
- Avoiding products that claim to prevent SIDS deaths;
- Avoiding over-reliance on heart, breathing, motion, or other monitors or alarms as a primary method of SIDS prevention;
- Avoiding swaddling once the baby starts to roll over;
- Giving the baby plenty of supervised "tummy time" when they are awake to help them strengthen their neck, shoulder, and arm muscles.

The AAP recommends against cosleeping (bedsharing) with infants for safety reasons, but many parents will choose this arrangement for a variety of reasons. A Canadian study found that one-third of mothers reported bedsharing regularly, and almost as many reported bedsharing occasionally. Those who choose to bedshare are advised to avoid:

- Smoking, drinking alcohol, using illicit drugs, or any medications that cause drowsiness;
- Sleeping on a soft surface such as a waterbed, air mattress, sofa, or armchair;
- Having loose bedding, pillows, and other objects in the bed;
- Putting the baby down on their stomach or side;
- Sharing the bed with other people (such as other children) or animals.[23]

The appendix provides additional resources that offer safer ways to cosleep with an infant for those who find this is the best arrangement for them.

PERSONAL AND WORKPLACE BOUNDARIES

Parents may feel the need to enact boundaries surrounding the care of their baby and how others interact with them as they navigate this challenging time. Although most individuals have good intentions, they often have their own ideas about how things should be done based on their own previous experiences, their level of understanding of the risks babies face, and their personal preferences. It can be very challenging to confront close friends and family who want to interact with the baby but who do not have the same standards regarding safety and childrearing as the parents. However, it is the parents' right and responsibility to make choices regarding the safety and well-being of their child and to insist that those who wish to be a part of their child's life respect their decisions.

Some common boundaries that many parents will feel the need to articulate include:

- Being vaccinated before interacting with the baby;
- Waiting until the parents are ready to receive visitors;
- Calling first before coming over;
- Washing hands before holding the baby;
- Not coming over if any individual in the family has recently shown signs of contagious disease;
- Not kissing the baby;
- Following the parents' guidance on feeding, including timing and storage/preparation of bottles;

- Following the parents' guidance regarding sleep, including safe sleep practices, timing, and location of naps;
- Being supportive or at least respectful of parents' decisions.

Some parents on leave may feel the need to enforce boundaries regarding workplace communication during their leave. Although each individual may differ in terms of how much contact they want to have during their leave, they should feel free to determine that they do not want any contact unless another arrangement has been agreed upon and they are being compensated for any work that they do.

KEEPING IN TOUCH DURING LEAVE

Some parents who are on leave will want a complete separation from the workplace during the entire time they are gone. This is perfectly reasonable in most cases and can help new parents focus on their own healing and bonding and caring for their new child. Other parents either need or want to be more connected and apprised of the activities in the workplace, even when they are not physically there. Practice owners, partners, and managers may want or need to participate in certain tasks during their leave such as managing the finances of the practice or fielding concerns and questions for employees. Even apart from ownership and managerial responsibilities, some parents may want to continue regular communication with their workplace to have updates on certain patients or projects and to receive support from friends and coworkers. Parents who take longer leaves sometimes worry that their time away from work will cause them to forget how to perform their job well and lose some of their skills. This worry can increase the anxiety around going back to work and may affect how much leave a parent feels comfortable taking.

In the UK, where maternity leave may last 26 weeks or more, employees on leave are permitted 10 paid workdays during their leave, called "keeping in touch" or KIT days, without having to end their leave. KIT days are optional and must be agreed upon by both the worker and the employer. These days are available to parents on maternity leave and adoption leave. In dual-income households where both parents are taking shared parental leave, the two parents can share 20 paid workdays, which are referred to as "shared parental leave in touch" or SPLIT days. These days will not affect an employee's eligibility for their full parental leave and pay. Working a small number of days over the course of parental leave, particularly a maternity leave of six months or more, can help parents feel more connected to their workplace and any changes that may have happened while they were on leave. It can provide needed coverage for the workplace, and it can allow parents on leave to help maintain their skills and their confidence during their time away from practice.[24]

TAKE-HOME POINTS

- No two fourth trimesters will be the same. Each parent and each baby will have their own distinct needs. Every family will benefit from community support;

- Postpartum health care is very important and can help prevent potentially life-threatening consequences. The US lags far behind other developed countries in the comprehensiveness of the care provided. It is worth taking advantage of every available resource and reaching out promptly with any concerns;
- Infant care includes establishing feeding, protection from infectious disease, practicing safe sleep habits, and much more. Parents should consult their healthcare providers with any questions or concerns;
- Boundaries around baby care and interaction may be difficult to put in place, but they are essential to help parents feel supported in the work they do to keep their child safe;
- Workplace boundaries may be necessary to allow for rest, healing, and bonding before returning to work.

REFERENCES

1. Columbia University Irving Medical Center. A mother's guide to the fourth trimester. November 9, 2021. Accessed November 30, 2022. https://www.cuimc.columbia.edu/news/mothers-guide-fourth-trimester

2. American College of Obstetricians and Gynecologists. Optimizing postpartum care. May 2018. Accessed December 7, 2022. https://www.acog.org/clinical/clinical-guidance/committee-opinion/articles/2018/05/optimizing-postpartum-care

3. National Institute for Health and Care Excellence. Postnatal care. April 20, 2021. Accessed December 21, 2022. https://www.nice.org.uk/guidance/ng194/chapter/Recommendations

4. Centers for Disease Control and Prevention. Maternal mortality. September 16, 2022. Accessed November 30, 2022. https://www.cdc.gov/reproductivehealth/maternal-mortality/index.html

5. National Health Service. Health visitor. Accessed December 7, 2022. https://myhealth.london.nhs.uk/maternity/ive-had-my-baby/health-visitor/

6. The Commonwealth Fund. Maternal mortality and maternity care in the United States compared to 10 other developed countries. November 18, 2020. Accessed December 27, 2022. https://www.commonwealthfund.org/publications/issue-briefs/2020/nov/maternal-mortality-maternity-care-us-compared-10-countries

7. Stanford Medicine Children's Health. The new mother: Taking care of yourself after birth. Accessed December 21, 2022. https://www.stanfordchildrens.org/en/topic/default?id=the-new-mother-taking-care-of-yourself-after-birth-90-P02693

8. Stanford Medicine Children's Health. The neonatal intensive care unit (NICU). Accessed December 21, 2022. https://www.stanfordchildrens.org/en/topic/default?id=the-neonatal-intensive-care-unit-nicu-90-P02389

9. March of Dimes. Coping with stress in the NICU. Accessed December 21, 2022. https://www.marchofdimes.org/find-support/topics/neonatal-intensive-care-unit-nicu/coping-stress-nicu

10. March of Dimes. Your baby's NICU stay. Accessed December 21, 2022. https://www.marchofdimes.org/find-support/topics/neonatal-intensive-care-unit-nicu/your-babys-nicu-stay

11. University of North Carolina at Chapel Hill. 4th trimester project. Accessed November 30, 2022. https://newmomhealth.com
12. National Health Service; Better Health, Start for Life. Breastfeeding challenges. Accessed December 25, 2022. https://www.nhs.uk/start4life/baby/feeding-your-baby/breastfeeding/breastfeeding-challenges/tongue-tie/
13. US Department of Agriculture; WIC Breastfeeding Support. Common breastfeeding challenges. Accessed December 25, 2022. https://wicbreastfeeding.fns.usda.gov/common-breastfeeding-challenges
14. The Lactation Network. Meeting with a lactation consultant. Accessed December 26, 2022. https://lactationnetwork.com/blog/meeting-with-a-lactation-consultant/
15. American Academy of Pediatrics; healthychildren.org. Formula feeding. Accessed December 27, 2022. https://www.healthychildren.org/English/ages-stages/baby/formula-feeding/Pages/default.aspx
16. Duda K. 7 ways to protect your baby from cold and flu. September 20, 2021. Accessed December 19, 2022. https://www.verywellhealth.com/ways-to-protect-your-baby-from-cold-and-flu-4057987
17. Cherry JD. Pertussis in young infants throughout the world. *Clin Infect Dis.* 2016;63(Suppl 4):S119–S122.
18. Tregoning JS, Schwarze J. Respiratory viral infections in infants: Causes, clinical symptoms, virology, and immunology. *Clin Microbiol Rev.* 2010;23(1):74–98.
19. Centers for Disease Control and Prevention. Whooping cough is deadly for babies. December 1, 2022. Accessed December 27, 2022. https://www.cdc.gov/pertussis/pregnant/mom/deadly-disease-for-baby.html
20. Centers for Disease Control and Prevention. Vaccines for family and caregivers. November 9, 2021. Accessed December 27, 2022. https://www.cdc.gov/vaccines/pregnancy/family-caregivers.html
21. Centers for Disease Control and Prevention. Sudden unexpected infant death and sudden infant death syndrome. June 21, 2022. Accessed December 16, 2022. https://www.cdc.gov/sids/data.htm#race
22. National Institutes of Health. Safe to sleep. Ways to reduce baby's risk. Accessed December 14, 2022. https://safetosleep.nichd.nih.gov/safesleepbasics/risk/reduce
23. Government of Canada. Joint statement of safe sleep: Reducing sudden infant deaths in Canada. October 8, 2021. Accessed April 25, 2023. https://www.canada.ca/en/public-health/services/health-promotion/childhood-adolescence/stages-childhood/infancy-birth-two-years/safe-sleep/joint-statement-on-safe-sleep.html#bs
24. GOV.UK (UK Government). Employee rights when on leave. Accessed December 16, 2022. https://www.gov.uk/employee-rights-when-on-leave

11

Mental health

During and after my first two pregnancies, I remember coping pretty well with all of the changes both inside of and around me, apart from some temporary baby blues. Eight years later, however, when I had my third child, I felt completely broken in the first few months. I cried endlessly. I begged my husband to come home from work. Even though I felt lost in the lack of structure of my time on maternity leave, I dreaded going back to work. I felt so much guilt for not being "happy." When I was back at work, the guilt just compounded. I worried that my baby was suffering because of me and that my children would be better off with someone else. I cried at work on my lunch break pretty much every day. Eventually, my sisters suggested that I seek treatment. My Ob/Gyn prescribed an antidepressant, and I found a therapist. Both of those things really helped. I stayed on antidepressants throughout my next pregnancy and figured I would be okay. However, I again experienced very similar signs, even on medication. I ended up needing a dose increase, and I went back to therapy. Soon, the despair and the hopelessness started to fade. I will forever be grateful to those who offered and recommended help and support.

 DOI: 10.1201/9781003406907-11

MENTAL HEALTH AND MENTAL WELLNESS

According to the US Department of Health and Human Services, mental health is defined as including "emotional, psychological, and social well-being." Multiple factors can negatively affect mental health, including biological factors (genes, illness elsewhere in the body, injury, and the chemistry of the brain), life experiences (a history of abuse or trauma), and a history of mental health disorders in other family members. A person's mental health will affect how they live and work, how they think and reason, how they cope with stress, and how they relate to others. In the year 2020, one in five adults was reported to have had a challenge with their mental health and one in six "young people" was reported to have experienced a major depressive episode.[1]

Although there is still much to learn about mental health and how to best support good mental well-being, it is known that an individual's mental health can change significantly throughout their lives, whether for the better or worse. Life experiences, including veterinary education and employment, can certainly play a role in the mental health of an individual. In addition, all of the changes that take place in a person's body during and after pregnancy, along with the physical and emotional demands of parenthood, can create significant challenges to an individual's emotional well-being.

MENTAL HEALTH RISK IN VETERINARY MEDICINE

LACK OF EVIDENCE FOR INCREASED PREDISPOSITION FOR POOR MENTAL HEALTH

Mental health and well-being are frequently discussed topics in veterinary medicine because they relate to increasing rates of burnout, leaving the profession, and in some cases, suicide. Burnout, defined as a "psychological syndrome emerging as a prolonged response to chronic interpersonal stressors on the job," is often cited as a primary cause for attrition of veterinary professionals in the face of increasing demand for their services. The main signs of workplace burnout include "overwhelming exhaustion, feelings of cynicism and detachment from the job, and a sense of ineffectiveness and lack of accomplishment."[2]

Poor mental health is also a major concern when considering the elevated rates of suicide among veterinary professionals. Many theories are offered about the cause of poor mental health and increased rates of suicide and suicidal ideation among some veterinary workers.[3] These include the high stress level and long work hours often associated with clinical practice, negative client interactions and cyberbullying, and presumptions that those who choose to enter the field of veterinary medicine are more likely to have personality traits that predispose them to mental illness. Although it has been suggested that veterinary students and veterinary professionals are more predisposed to possessing "maladaptive" personality traits than those of other healthcare professions, this has not been proven in studies.

One UK study compared 1744 students (80% of whom were female) in various professional degree programs (veterinary, medicine, pharmacy, dentistry, and law) by means of an online questionnaire. The students were scored on the "big five" personality traits: neuroticism, extraversion, openness, agreeableness, and conscientiousness. Their levels of perfectionism, well-being, psychological distress, depression, and suicidal ideation or attempts were also measured. Law students were chosen to participate alongside students in healthcare professions because they were determined to have similarly stressful schedules and workloads. However, their need to display empathy was considered to be much lower than those in healthcare professions, as law is said to demand more "objectivity" and a "dispassionate approach."[4]

The results of this survey found few statistically significant differences between groups of professional students in the big five personality categories. Veterinary students were significantly more open to new experiences than pharmacy students and were significantly more agreeable (along with medical and dental students) than law students. These traits were both associated with more positive mental health. No significant differences were measured between groups of students on the traits of neuroticism, extraversion, or conscientiousness. The researchers concluded that, although no significant differences could be measured between the various groups of professional students that would cause a predisposition for poor mental health, within each group, individuals who scored higher on neuroticism and lower on conscientiousness were at greatest risk for mental health challenges.[4]

Studies have also been done to compare veterinary students with the general adult population to look for any predisposing causes for poor mental health. Rates of adverse childhood experiences (ACEs) were measured in veterinary students in the United States and compared with their rates of depression and stress. They also evaluated the age at which each student decided they wanted to become a veterinarian in relation to the time ACEs were experienced. ACEs were categorized as abuse, neglect, or household dysfunction experienced before the age of 18. The question asked in this study was: "Do people come into the veterinary profession with a predisposition to poor mental health or does poor mental health arise as a result of academic veterinary medical training and the practice of veterinary medicine?"[5]

The results indicated that the rates of ACEs reported by veterinary students did not differ from the rates reported by the general adult population in the United States. No relationship was found between the age at which a student first wanted to become a veterinarian and any exposure to ACEs. Even without an increased exposure to ACEs, 43% of students surveyed met the criteria for depression, and 54% of students surveyed met the criteria for above-average stress levels. The researchers concluded that veterinary students did have an increased risk for "poor mental health outcomes," but that exposure to ACEs alone did not appear to be the cause for the increased risk.[5]

Although ACEs alone cannot explain the increased risk of depression and mental health issues for veterinary students and professionals, they may still be part of the picture. Veterinary students did experience ACEs at rates in line with the general US adult population. Coupled with the "unique stressors in veterinary medicine," this may explain some of the increased risk. For example, programs for medical students have been described as creating a "hidden curriculum" that encourages unhealthy practices such as working long hours, not asking for help, not having personal time, feeling ostracized for making mistakes, and not displaying or being aware of one's emotions. The authors of the study suggested that some veterinary curricula likely reinforce similar messages. Unlike the practice of medicine for humans, however, veterinarians are more likely to be saddled with a higher debt-to-income ratio and have the added challenge of having to consider the financial limitations of their clients and sometimes make decisions to euthanize based on finances.[5] Economic euthanasias don't just negatively affect veterinarians, but the entire veterinary team. This can be compounded by angry reactions from clients who direct their frustration toward the veterinary team.

ENVIRONMENTAL FACTORS FOR POOR MENTAL HEALTH IN VETERINARY PROFESSIONALS

The 2021 Merck Veterinary Wellness Study measured rates of stress, including burnout, serious psychological distress, and suicidal ideation among veterinarians and veterinary staff. They found the most important reasons cited for severe psychological distress to be staff shortages, student debt, the COVID-19 pandemic, long work hours, and high reported suicide rates among veterinarians. The percentage of veterinary professionals reporting serious psychological distress levels had increased from 5.3% in 2017 and 6.4% in 2019, to 9.7% in 2021. The well-being of veterinary staff was significantly worse (almost twice as high) than that of veterinarians in this survey. Suicidal ideation was increased in veterinary support staff compared with the last well-being study conducted in 2019.[6]

Groups with different racial and ethnic identities had different unique stressors and stress levels as well. For example, Hispanic, Black, and Asian respondents listed the lack of diversity within the veterinary profession as the highest on their list of stressors. Serious psychological distress rates were also higher in Black and Hispanic veterinarians when compared with White veterinarians. When the groups were separated by age, serious psychological distress was highest in the youngest group of veterinarians. Although less than 5% of veterinarians under the age of 55 reported being likely to leave the veterinary profession within the next two years, almost one-third of veterinary staff reported being likely to leave veterinary practice within the next two years.[6] An unrelated study of job stressors and employee health also found that workers who reported negative work-home interference (where their job negatively affects their life outside of work), especially mothers, were at higher risk for anxiety and depression.[7]

PERINATAL MENTAL HEALTH DISORDERS IN VETERINARY PROFESSIONALS

There is little data on reported rates of perinatal mental health disorders in veterinarians and other veterinary professionals. A study by Wayne et al. sampled 1082 veterinarians who identified as mothers (or were pregnant) in the United States by way of an online questionnaire. One hundred eighty-one respondents (16.7%) reported having been diagnosed with postpartum depression, and another 353 respondents (32.6%) reported signs consistent with postpartum depression but had not sought any treatment. These rates may not be truly representative of the population, because respondents who experienced poor mental health may have been more likely to respond to the questionnaire than those who had not. However, the rates are still significantly higher than average rates for perinatal mental health disorders (to be discussed in the next section).[8,9]

PERINATAL MENTAL HEALTH DISORDERS

Postpartum Support International defines perinatal mental health disorders as mood disorders that develop either during pregnancy or in the first 12 months after giving birth. These disorders are said to affect about 15–20% of pregnant people. The most well-known is postpartum depression, but this is just one disorder in the category. Others include anxiety, obsessive–compulsive disorder (OCD), post-traumatic stress disorder, bipolar mood disorders, and psychosis.[10]

PERINATAL DEPRESSION

Depression can develop at any time during pregnancy or after delivery. Although the exact causes are still unknown, having a previous history of depression is the greatest predictor of its development. Other suspected contributing factors include hormonal changes, lack of sleep, and genetics. It occurs in about 10% of pregnant people and in 15% of postpartum mothers.[11,12]

Some postpartum parents will experience a transient form of mild depression, often called "baby blues," that typically lasts no longer than 2 weeks and does not interfere with their ability to care for themselves or their child. Signs of baby blues can include:

- Mood swings;
- Irritability;
- Anxiety;
- Decreased concentration;
- Trouble sleeping;
- Excessive crying.

These signs are all temporary and resolve on their own. When they do not resolve or they become more severe, postpartum depression may be diagnosed.

Postpartum Support International and the *Merck Manual* report that there are many possible risk factors for perinatal depression, including:

- "Baby blues";
- Poverty;
- Being a teenager at the time of pregnancy;
- Family history of depression;
- Previous personal history of depression or postpartum depression;
- Experiencing one or more major life stressors;
- Lack of family support;
- History of previous adverse birth outcomes;
- Complications during a current pregnancy or delivery;
- Being pregnant with or having delivered multiples;
- Having had fertility treatments;
- Having a baby that spends/spent time in the neonatal intensive care unit (NICU);[13]
- History of mood changes associated with menstrual cycle or contraceptive use;
- Ambivalence about a current pregnancy;

- Thyroid imbalance;
- Diabetes mellitus (types 1 or 2, or gestational);
- Problems with breastfeeding.

Signs and symptoms of perinatal depression can include:

- Extreme sadness;
- Mood swings;
- Uncontrollable crying;
- Trouble sleeping;
- Decreased appetite or overeating;
- Irritability;
- Anger;
- Headaches;
- Body aches;
- Extreme fatigue;
- Irrational worries about the baby or the ability to care for the baby;
- Fear of harming the baby;
- Disinterest in the baby;
- Guilt;
- Suicidal ideation;
- Anxiety;
- Panic attacks.

These changes often appear gradually over about three months, but they can begin more suddenly. In whatever form they appear, these mental changes can interfere with a mother's ability to care for themselves and/or their child. Although some cases of perinatal depression may resolve without treatment, some will turn into chronic depression. About one-fourth to one-third of cases recur at some point throughout life.[14,15]

PERINATAL ANXIETY

According to Postpartum Support International and the MGH Center for Women's Mental Health, perinatal anxiety is said to develop in about 6% of pregnant people and 10% of postpartum people. Although it can be experienced *without* depression, it is often experienced with depression. The risk factors for anxiety include:

- Personal or family history of anxiety;
- History of perinatal anxiety or depression;
- Thyroid imbalance.

Symptoms of perinatal anxiety include:

- Frequent or constant worrying;
- Racing thoughts;
- Trouble eating;
- Trouble sleeping;
- Dizziness;
- Hot flashes;

- Nausea;
- Restlessness;
- Thoughts that something bad will happen.

It can sometimes be hard to tell the difference between postpartum depression and postpartum anxiety, because many of the symptoms overlap, and also because anxiety is listed as a symptom of depression. Postpartum anxiety is typically temporary and tends to lessen over time, but it is treatable with therapy and/or medication.[16,17] Within the anxiety diagnosis are additional more specific diagnoses, including postpartum panic disorder and postpartum OCD.[17]

BOX 11.1   Practice profile: Postpartum depression and anxiety

Marie Holowaychuk, DVM, Dipl. ACVECC

As a specialist veterinarian, I thought completing a residency in emergency and critical care would be the hardest thing I would do in my life. That was until I became a mom. I chose to become a single mom in 2018 and after 2 years of intermittent attempts at pregnancy, I had my daughter in June of 2020. Having spoken to a psychologist and other mental health professionals leading up to my pregnancy, I knew that being a single mom would require additional supports and would be exceptionally difficult. What I didn't realize was just how difficult it would be.

As someone who has lived with anxiety and depression since my early 20s, I also knew that I was at additional risk of postpartum mental illness. So, it's not surprising that amid the pandemic, the isolation experienced by most mothers postpartum, and the insomnia, unpredictability, and uncertainty that occur during the first several months of a baby's life, I developed severe postpartum anxiety and depression.

At first, I thought my constant crying was normal and that all new moms felt anxious as they navigated their baby's first few months of life. However, when I began to have panic attacks, moments of rage, and periods of intense sadness, I knew I needed help. And thankfully, the nurse who administered my daughter's 8-week vaccines was equally invested in my mental health and insisted that I speak to my physician. So, I made the appointment and shared what I was going through, and right away my diagnosis of postpartum depression and anxiety was confirmed. I started taking a medication that was safe for breastfeeding and I began to seek additional support from a postpartum counselor and my naturopathic doctor. As the medication took effect and with the additive support of mental health appointments and nutritional supplements, I improved.

And when my daughter was approximately 3 months old, I was able to get back to regular exercise, my sleep became more consolidated again, and I resumed prioritizing self-care in the form of time outside, virtual connections with friends, and time off just for me while a nanny or babysitter took over. At first it felt self-ish and unnecessary to do these things for myself. But then I was reminded of my advocacy in veterinary mental health and well-being and accepted that self-care for caregiving moms is equally as important as self-care for caregiving veterinary professionals.

My daughter recently turned 2 years old, and I can confidently say that my mental and physical health are the best they have ever been. I am in the process of weaning from my postpartum medication and know that my prioritization of boundaries, saying no, and work-life separation will support me through this transition. My daughter has brought immense joy and love into my life, and regularly holds me accountable for my self-care in the form of an early bedtime, as well as daily exercise, outdoor time, and meditation. And I could not be more grateful.

POSTPARTUM PANIC DISORDER

Postpartum panic disorder causes intense nervousness and panic attacks. The most common signs of a panic attack include:

- Shortness of breath;
- Chest pain;
- Heart palpitations;
- Claustrophobia;
- Dizziness;
- Numbness and tingling in one or more limbs.[17]

PERINATAL OBSESSIVE–COMPULSIVE DISORDER

A poorly known and often misdiagnosed perinatal mood disorder, OCD can be diagnosed during pregnancy, postpartum, and in some new fathers (and other non-birthing parents). Postpartum Support International reports that about 3–5% of new parents can experience symptoms of OCD, which can include intrusive thoughts or obsessions that are persistent and related to the baby (these can be upsetting and may be related to thoughts of the baby experiencing harm). Parents may try to cope with these horrific thoughts through compulsive behavior, where they repeat certain actions in an attempt to protect the baby or allay their fears. Parents with OCD are usually aware that their thoughts are abnormal, and they are not delusional. They are not likely to act on any of the thoughts that they have, other than to make efforts to prevent any perceived harm to their child.[18]

BIPOLAR MOOD DISORDERS

Bipolar mood disorder types 1 and 2 cause individuals to experience emotional highs (mania) followed by emotional lows (depression). Persons with bipolar disorder type 2 typically have fewer manic episodes and less extreme swings between emotional highs and lows. The criteria for these disorders include symptoms that "last longer than four days and interfere with functioning and relationships." Risk factors for bipolar mood disorder include having a relative with a bipolar mood disorder. Some people will have already been diagnosed with bipolar disorder before pregnancy, but others will first see symptoms appear during pregnancy or the postpartum period.[19]

Signs of bipolar disorder can include periods of depression and irritability that may oscillate with periods of mania. Mania can include racing thoughts, trouble concentrating, little need for sleep, impulsiveness, overconfidence, grandiose thoughts, high energy, rapid speech, and significantly increased mood. Some individuals can experience psychosis as part of either the mania or the depression, and this can include having hallucinations or delusions. These require urgent treatment.[19]

It is very important to seek help from a qualified professional who can determine whether depression is also accompanied by mania. If only the depression is treated, there is a high risk of exacerbating the manic side of the disorder.[19]

POSTPARTUM PSYCHOSIS

Postpartum psychosis occurs in about 0.1–0.2% of births. Although it can develop any time during the first year after delivery, it most commonly presents suddenly, within the first two weeks. Risk factors include a previous history (or family history) of a bipolar disorder or a previous psychotic episode. Symptoms of psychosis include:

- Delusions;
- Hallucinations (this can include both seeing things and hearing things);
- Irritability;
- Hyperactivity;
- Inability to sleep or decreased need for sleep;
- Paranoia;
- Rapid mood swings;
- Difficulty communicating.

Although most affected mothers do not experience any kind of violent delusions, there is an approximately 5% suicide rate and a 4% infanticide rate associated with postpartum psychosis. The breaks from reality a mother experiences during psychosis can make thoughts of violent actions seem sensible. Even people who are not experiencing any kind of violent delusions can experience irrational judgment, which can endanger themselves and their families. For these reasons, emergency care is essential for any suspected case of psychosis.[20]

POSTPARTUM POST-TRAUMATIC STRESS DISORDER

About 9% of people experience postpartum post-traumatic stress disorder (PTSD). This is most commonly attributed to a "real or perceived trauma during delivery or postpartum" such as:

- An emergency cesarean delivery;
- Complications during delivery;

- Use of a vacuum extractor or forceps to deliver the baby;
- Postpartum hemorrhage;
- Perineal tears;
- Unexpected hysterectomy;
- Preeclampsia;
- Heart disease;
- Baby requiring NICU care.

In some cases, the trauma is less physical and more emotional in the form of "feelings of powerlessness, poor communication and/or lack of support and reassurance during delivery." A history of previous trauma, such as rape or sexual abuse, can also increase the risk of postpartum PTSD.[21]

Symptoms associated with this PTSD can include:

- Flashbacks;
- Nightmares;
- Anxiety;
- Panic attacks;
- Irritability;
- Difficulty sleeping;
- Hypervigilance;
- Exaggerated startle response;
- A sense of detachment.

Some mothers may feel a desire to avoid anything that reminds them of the traumatic event, including avoidance of people and places.[21]

WAYS TO GET HELP

Whether it is related to a perinatal mental health disorder or not, any individual who is experiencing suicidal ideations should seek professional help. Depending on the urgency, this may include consulting with a mental health professional or other healthcare provider. In an emergency, local emergency services (911 in the US) and and/or a suicide and crisis hotline (988 in the US) should be contacted immediately. The appendix lists some contact resources for other countries.

When emergency assistance is not needed, there are many resources available to individuals experiencing mental health disorders. Although some disorders may resolve without treatment, some can be chronic or can have life-threatening consequences; thus, it is always best to seek help. There are a variety of options available to pregnant people and parents who need support, and there are some that are specifically meant for veterinary professionals. The following forms of treatment and resources are not meant to be exhaustive, nor are any of them suggested to be singular treatments. It is often the case that more than one of these will be needed to provide relief to the affected parent.

Both the American College of Obstetricians and Gynecologists and the American Academy of Pediatrics recommend that pregnant people and postpartum mothers be screened for perinatal depression during pregnancy, at postpartum visits, and at the 1-,

2-, 4-, and 6-month well-child visits.[22,23] In the UK, the National Health Service recommends that healthcare providers and other practitioners who visit and care for the parents and baby at home starting shortly after birth should also screen for postnatal depression.[24] Healthcare providers who determine that a patient has a perinatal mood disorder can then help to coordinate care if they are not the ones to provide it themselves. Non-birthing parents can also experience many of these mental health disorders during the pregnancy and postpartum periods, and they should watch for any of the above-mentioned symptoms and seek treatment if needed.

MEDICATION

The Centre for Addiction and Mental Health (CAMH) reports that "treatment with both medication and psychotherapy is often necessary for moderate to severe perinatal symptomatology." Medication is usually required for individuals who find it difficult to care for themselves or their baby, or if they have any suicidal ideations.[25]

Healthcare providers including physicians (obstetrician-gynecologists, general practitioners, and psychiatrists), nurse practitioners, physician assistants, some midwives, and other providers with the authority to prescribe may be able to prescribe antidepressants and other medication to treat perinatal mood disorders. Some practitioners may not feel comfortable diagnosing or treating one or more of these conditions, especially if they are concerned about bipolar disorder or psychosis, or if it is not something they treat regularly. They may instead refer a patient to a psychiatrist or other provider who specializes in treating people with perinatal mood disorders.

There are multiple types of medications used to treat perinatal mood disorders. Healthline reports that antidepressants are the most common type, and they can include selective serotonin reuptake inhibitors like fluoxetine and sertraline, "atypical" antidepressants like bupropion (a dopamine reuptake blocker), and less commonly, tricyclic antidepressants (like amitriptyline) or monoamine oxidase inhibitors (like selegiline). All of these medications work to balance chemicals in the brain to help relieve symptoms of depression. All of them also carry some risk of side effects, which can include sedation, weight gain, nausea, blurry vision, constipation, and sexual problems.[26]

There is only one medication that is FDA approved for the treatment of postpartum depression. According to the patient brochure, Zulresso (brexanolone) is given by continuous intravenous infusion over 60 hours and may provide relief from symptoms in that amount of time. Side effects can include excessive sedation, facial flushing, loss of consciousness, and suicidal thoughts. It is typically considered for patients who have not responded well to other medications.[27]

Other classes of medication, including anxiolytics, antipsychotics, and lithium compounds are often used to treat postpartum psychosis, usually when the patient in still in the hospital.[28] Lithium, antipsychotics, and anticonvulsants are often used to treat perinatal bipolar disorders.[29]

Mothers who are pregnant or breastfeeding may be hesitant to take medication out of concern for any negative effects the medication could have on their child. In addition to consulting with their provider to understand the risks and benefits of medication, Postpartum Support International offers other resources, including Mother to Baby, a service "dedicated to providing evidence-based information to mothers, healthcare professionals, and the general public about medications and other exposures during pregnancy and while breastfeeding"; and the Infant Risk Center, "a call center based solely on evidence-based medicine and research … dedicated to providing current and accurate information to pregnant and breastfeeding mothers and healthcare professionals." The Infant Risk Center also trains pharmacy and medical students on the safe use of medication during pregnancy and lactation.[30]

Psychotherapy

According to the National Institute of Mental Health, mental health counselors can provide psychotherapy that can help treat a variety of perinatal mental health disorders, often in combination with medication. Depending on the therapist, the patient, and the mood disorder, different types of therapy may be explored.

Common evidence-based therapies include cognitive behavioral therapy (CBT) and interpersonal therapy (IPT). CBT can be used to treat individuals with depression and/or anxiety, or other conditions, to change the way they think, behave, and/or react to stressors. CBT can help people learn to reframe their thoughts and behaviors in healthier ways and challenge unhealthy thinking and behavior patterns. This can be done in a one-on-one setting, or in a support group setting. IPT focuses on improving communication skills, developing a support system, and adjusting expectations to be more realistic to help individuals with depression cope more effectively.[31]

Postpartum PTSD can also be treated with psychotherapy, but an additional treatment option that is often used to treat unprocessed trauma has also been shown to help parents with postpartum PTSD in their healing. Eye movement desensitization and reprocessing (EMDR) is another evidence-based treatment that helps the body process traumatic experiences in which the individual seems to be "stuck." This therapy is provided by a trained mental health counselor. According to the EMDR International Association, the process involves revisiting the traumatic experience while using bilateral stimulation (either continuous stimulation of both arms or hands, or bilateral eye movements) to desensitize the body to the triggers of the traumatic experience. EMDR has been proven to reduce the "vividness and emotion associated with trauma memories." It has also been successfully used in the treatment of anxiety, depression, and OCD.[32,33]

Support groups

Postpartum Support International reports that support groups can be an important part of treatment for a variety of perinatal mental health disorders. These groups can meet in person or be attended virtually, and they can allow participants to hear the stories of others navigating similar challenges. Some support groups are specific to certain disorders (e.g., bipolar disorder), identities (e.g., Black moms, military moms), or traumatic experiences (e.g., birth trauma, NICU stay). Those who have experienced a loss, whether through pregnancy termination, miscarriage, stillbirth, or infant death, can also find comfort and community through a support group. Hospitals, mental health counselors, and local, national, and international organizations all offer support groups specific to perinatal mental health disorders.[34] Support groups can be found through Postpartum

Support International (see the appendix), through local hospitals and mom groups, and through local mental health providers.

The benefits of support groups include having the support and experience of others and the opportunity to form a community with them while still having the guidance and supervision of a trained facilitator. Group therapy also tends to be significantly less expensive than individual therapy sessions, and in some cases, there is no cost at all. Some individuals will take part in a support group in addition to private therapy sessions. Disadvantages of a support group include less personal attention, the potential to be re-traumatized by the stories of others, less privacy, and the potential for conflict with others.[34]

COMMUNITY AND ONLINE SUPPORT

Apart from groups meant to support individuals with perinatal mental health disorders, other types of community groups allow a space for new parents to meet and share their challenges and successes with each other. Some parents will find support through their church congregation, gyms or sports activities, coworkers, neighbors, school parenting groups, and other communities. Different online and social media groups exist for parents to find others who are in similar life stages, where they can provide and receive support, ask questions, and share relatable experiences. These include the private Facebook group DVMoms-Life in the Trenches, which currently has more than 17,000 members from the US and around the world, and the group's many private Facebook offshoot groups. Other Facebook groups include Vet Mums based out of the UK and Vet Tech Moms. Although these groups do not have any formal mission of supporting veterinary professionals through mental health challenges, they can be spaces to form friendships, share experiences, and feel less alone. Disadvantages of online groups can be similar to those of support groups, including receiving unsolicited advice, loss of privacy, the tendency to compare to others, and conflict over differences in opinion.

Any support that others can provide that reduces the physical and mental load on parents of infants and young children can be extremely beneficial. This can include organized meal trains (including digital restaurant gift cards or meals ordered through a meal delivery service), grocery delivery, meal subscription services, housekeeping and cleaning services, postpartum doula services, help with care and transportation of children, and help with any other personal or family needs. Parents who are navigating perinatal mood disorders may also benefit from having those around them listen to their concerns, validate their feelings, and recommend professional help if none has been accessed.

VETERINARY RESOURCES

Veterinary professionals also have resources that are specific to their unique mental health needs. Although they are not specifically tied to perinatal mental health disorders and do not provide any individual therapy or treatment, they can still be a valuable resource to those who need support.

Not One More Vet (NOMV) is a charity founded by a veterinarian in 2014 in the wake of the suicide of a veterinary colleague, with the hope that not one more vet would die by suicide. According to their website, "NOMV helps veterinary professionals through peer-to-peer support, financial support grants, education presentations, and by collaborating with partner agencies to extend services to the veterinary community." The website offers crisis helpline contact information by country and also offers Lifeboat, a "peer-to-peer support and mentorship" program that pairs veterinary professionals in need with a trained volunteer (who is also a veterinary professional) to communicate with them asynchronously for support. NOMV also offers CLEAR Blueprint, a practice certification program developed to help teach veterinary practices how to offer a healthy culture to their employees.[35]

Not Another Vet Nurse is an organization that provides mental health resources and support for veterinary support staff worldwide. They also help to bring awareness to the increased risk of suicide among veterinary support professionals and offer suicide prevention resources. Members can join forums and share their experiences with others, look for a new job, and send messages of encouragement to others in the field.[36]

The Veterinary Hope Foundation (VHF) was founded in 2021 by two veterinarians, also in response to reports of veterinary professionals dying by suicide. According to VHF's website, their mission is to support the mental health of veterinary professionals. Small online support groups with a veterinary focus are organized and led by mental health professionals. Each group focuses on a particular need in the community, such as grief and loss, the needs of associate veterinarians, owner and manager veterinarians, and all veterinary professionals.[37]

The VIN Foundation offers Vets4Vets and Support4Support, for veterinarians and support staff, respectively, at no charge and without any kind of membership required. Support can be in the form of one-one-one meetings with a veterinary professional who is trained as a mental health supporter, online weekly group meetings, substance abuse and behavioral addiction recovery groups, groups for neurodivergent veterinary professionals, and groups for those with a cancer diagnosis. These programs can help with navigating both work-related and non-work-related stressors.[38]

SELF-CARE

Self-care is often suggested as a remedy for both physical and mental health conditions, as well as for overwhelm, burnout, and many other life stressors. This can be a hard concept to embrace, because it tends to place the onus for healing on the individual who is suffering and often already overwhelmed to do more work to care for themselves. Martinez et al. reviewed many different definitions and concepts of self-care. Among them are "fundamental principles of self-care" as defined by the World Health Organization. These include "autonomy, self-efficacy, empowerment, community involvement, and community empowerment."[39] This definition does not place all the responsibility on the individual but rather emphasizes the individual's power and involvement in their community and their need for community support.

There are important steps that parents facing mental health challenges can take, in addition to receiving treatment from qualified professionals and support from their community, to help improve and maintain their mental well-being. According to the National Institutes of Mental Health, these include:

- Exercise: 30 minutes of daily exercise, which can be spread out into smaller increments throughout the day, can improve health and mood;
- Regular, healthy meals: these provide energy and can improve mental clarity;
- Hydration: this includes limiting caffeine and soft drinks that can cause dehydration and drinking plenty of water;
- Developing a sleep schedule and making it a priority (as much as pregnancy and caring for children will permit);
- Relaxing activities: meditation, breathing exercises, journaling, or other calming activities. Insight timer, a meditation app, offers meditations specifically for veterinary professionals;[40]
- Setting priorities and boundaries: prioritizing rest and other basic needs, saying no to activities that are not essential, and not creating unrealistic expectations;
- Gratitude: making a habit of identifying specific things to be grateful for each day;
- Reframing negative thoughts: this does not mean ignoring or glossing over negative or harmful things that are happening, but trying to be more kind and compassionate and less judgmental of oneself;
- Social connection: staying in contact with family, friends, coworkers, and others who can provide support both physically and emotionally.[41]

When self-care is framed as a conservation of resources and a nurturing of the mind and body, it can feel less overwhelming and more approachable. Also, since veterinary professionals are driven to care for others, self-care serves as a way to "refill their cup" so that they can continue to do so.

TAKE-HOME POINTS

- Working (or studying) in veterinary medicine during pregnancy and the postpartum period can create significant mental health challenges;
- Other causes of perinatal mental health disorders include genetics and a history of previous mental health disorders;
- Perinatal mental health disorders are treatable. Different treatment options exist, and many types of healthcare providers can recommend and explain the various options;
- Those who are working through perinatal mental health disorders should lean on their community as much as possible for support, both emotionally and physically;
- No matter how isolating having a mental health issue can feel, there are others who understand and want to help. Connecting with others can be an important part of recovery;
- Self-care is not a replacement for professional help, but rather a way to supplement it and maintain mental well-being;
- You are not alone. You are loved and needed. You do not need to walk through this alone. You are the parent that your child needs. Your veterinary colleagues want you to feel supported and valued.

REFERENCES

1. MentalHealth.gov; US Department of Health and Human Services. What is mental health? February 28, 2022. Accessed December 28, 2022. https://www.mental-health.gov/basics/what-is-mental-health

2. Maslach C, Leiter MP. Understanding the burnout experience: Recent research and its implications for psychiatry. *World Psychiatry.* 2016 Jun;15(2):103–111.

3. Tomasi S, Fechter-Leggett E, Edwards N, Feddish A, Crosby A, Nett R. Suicide among veterinarians in the United States from 1979 through 2015. *J Am Vet Med Assoc.* 2019;254:1.

4. Lewis EG, Cardwell JM. The big five personality traits, perfectionism and their association with mental health among UK students on professional degree programmes. *BMC Psychol.* 2020;8:54. https://doi.org/10.1186/s40359-020-00423-3

5. Strand EB, Brandt J, Rogers K, Fonken L, Chun R, Conlon P, Lord L. Adverse childhood experiences among veterinary medical students: A multi-site study. *J Vet Med Educ.* 2017;44(2):260–267. doi: 10.3138/jvme.0816-123R

6. Merck Animal Health. Veterinary Wellbeing. Veterinary mental health and wellbeing and how to improve them: Learnings from the Merck animal health veterinarian wellbeing study III. January 2022. Accessed June 9, 2022. https://www.merck-animal-health-usa.com/about-us/veterinary-wellbeing-study

7. Peeters, MCW, de Jonge J, Janssen PMP, van der Linden S. Work-home interference, job stressors, and employee health in a longitudinal perspective. *International Journal of Stress Management* 2004;11(4):305–322.

8. Centers for Disease Control and Prevention. Depression among women. May 23, 2022. Accessed December 28, 2022. https://www.cdc.gov/reproductivehealth/depression/index.htm

9. Wayne AS, Mueller MK, Rosenbaum M. Perceptions of maternal discrimination and pregnancy/postpartum experiences among veterinary mothers. *Front Vet Sci.* 2020;7:91.

10. Postpartum Support International. Perinatal mental health disorders. Accessed December 28, 2022. https://www.postpartum.net/learn-more/

11. Centers for Disease Control and Prevention. Depression during and after pregnancy. April 29, 2022. Accessed December 28, 2022. https://www.cdc.gov/reproductivehealth/features/maternal-depression/index.html

12. Centers for Disease Control and Prevention. Identifying maternal depression. May 2, 2022. Accessed December 28, 2022. https://www.cdc.gov/reproductivehealth/vital-signs/identifying-maternal-depression/index.html

13. Garfield CF, Lee YS, Warner-Shifflett L, Christie R, Jackson KL, Miller E. Maternal and paternal depression symptoms during NICU stay and transition home. *Pediatrics.* 2021;148(2):e2020042747. doi: 10.1542/peds.2020-042747

14. Moldenhauer JS. Postpartum depression. September 2022. Accessed December 28, 2022. https://www.merckmanuals.com/professional/gynecology-and-obstetrics/postpartum-care-and-associated-disorders/postpartum-depression

15. Postpartum Support International. Depression during pregnancy and postpartum. Accessed December 28, 2022. https://www.postpartum.net/learn-more/depression/

16. MGH Center for Women's Mental Health. Is it postpartum depression or post-partum anxiety? What's the difference? September 30, 2015. Accessed December 28, 2022. https://womensmentalhealth.org/posts/is-it-postpartum-depression-or-postpartum-anxiety-whats-the-difference/

17. Postpartum Support International. Anxiety during pregnancy and postpartum. Accessed December 28, 2022. https://www.postpartum.net/learn-more/anxiety/

18. Postpartum Support International. Pregnancy or postpartum obsessive symptoms. Accessed December 28, 2022. https://www.postpartum.net/learn-more/obsessive-symptoms/

19. Postpartum Support International. Bipolar mood disorders. Accessed December 28, 2022. https://www.postpartum.net/learn-more/bipolar-mood-disorders/

20. Postpartum Support International. Postpartum psychosis. Accessed December 28, 2022. https://www.postpartum.net/learn-more/postpartum-psychosis/

21. Postpartum Support International. Postpartum post-traumatic stress disorder. Accessed December 28, 2022. https://www.postpartum.net/learn-more/postpartum-post-traumatic-stress-disorder/

22. The American College of Obstetricians and Gynecologists. Screening for perinatal depression. November 2018. Accessed January 23, 2023. https://www.acog.org/clinical/clinical-guidance/committee-opinion/articles/2018/11/screening-for-perinatal-depression

23. Earls M, Yogman M, Mattson G, Rafferty J, et al; Committee on Psychosocial Aspects of Child and Family Health. Incorporating recognition and management of perinatal depression into pediatric practice. *Pediatrics*. 2019;143(1):e20183259.

24. National Health Service. Health visitor. Accessed January 23, 2023. https://myhealth.london.nhs.uk/maternity/ive-had-my-baby/health-visitor/

25. The Centre for Addiction and Mental Health. Perinatal mood and anxiety disorders: Pharmacotherapy. Accessed January 23, 2023. https://www.camh.ca/en/professionals/treating-conditions-and-disorders/perinatal-mood-and-anxiety-disorders/perinatal-mood---treatment/perinatal-mood---pharmacotherapy

26. Healthline. What medications help treat depression? November 9, 2021. Accessed January 23, 2023. https://www.healthline.com/health/depression/medication-list#types

27. Sage Therapeutics. Zulresso patient brochure. Accessed January 23, 2023. https://www.zulresso.com/content/dam/www_zulresso_com-aem/images/content/ZULRESSO%20Patient%20Brochure.pdf

28. Osborne LM. Recognizing and managing postpartum psychosis: A clinical guide for obstetric providers. *Obstet Gynecol Clin North Am*. 2018;45(3):455–468.

29. Clark CT, Wisner KL. Treatment of peripartum bipolar disorder. *Obstet Gynecol Clin North Am*. 2018;45(3):403–417.

30. Postpartum Support International. Medication resources. Accessed January 23, 2023. https://www.postpartum.net/resources/medication-resources/

31. National Institute of Mental Health; National Institutes of Health. Perinatal depression. Accessed January 25, 2023. https://www.nimh.nih.gov/health/publications/perinatal-depression#part_6550

32. Chiorino V, Cattaneo MC, Macchi EA, et al. The EMDR recent birth trauma protocol: A pilot randomised clinical trial after traumatic childbirth. *Psychol Health*. 2020;35(7):795–810.

33. EMDR International Association. About EMDR therapy. Accessed January 26, 2023. https://www.emdria.org/about-emdr-therapy/

34. Postpartum Support International. PSI online support meetings. Accessed January 26, 2023. https://www.postpartum.net/get-help/psi-online-support-meetings/

35. Not One More Vet. Home page. Accessed January 26, 2023. https://www.nomv.org

36. Not Another Vet Nurse, Inc. Home page. Accessed April 18, 2023. https://notanothervetnurse.wixsite.com/navn

37. Veterinary Hope Foundation. Home page. Accessed January 26, 2023. https://veterinaryhope.org

38. VIN Foundation. Vets4vets. Accessed April 17, 2023. https://vinfoundation.org/resources/vets4vets/

39. Martínez N, Connelly CD, Pérez A, Calero P. Self-care: A concept analysis. *Int J Nurs Sci*. 2021;8(4):418–425.

40. Insight Timer. Vet meditation: Goal visualization. Accessed April 17, 2023 https://insighttimer.com/mindbodythrive/guided-meditations/vet-meditation-goal-visualization

41. National Institute of Mental Health; National Institutes of Health. Caring for your mental health. Accessed January 26, 2023. https://www.nimh.nih.gov/health/topics/caring-for-your-mental-health#part_8445

12

Returning to work

For all of my pregnancies, going back to work and being separated from my baby (and other children) has been agonizing, gut-wrenching, and torturous—and that was just the anticipation! Having spent months devoted to healing and caring for my new baby, I had no desire to be separated from them and to have them spend large portions of their day in someone else's care. I worried about how they would eat, how they would sleep, if I would miss important milestones in their development, and if they would even know that I loved them. Each time I went back, there was a period of time where the sting of having to leave them was so great that I dreaded it. With time, it became less intense. I had to be told repeatedly by others that I was a great mom, that my children were safe, and that they knew that I loved them. I did enjoy being able to have more structure to my day, to interact with others who didn't need me every second of the day the way my baby did, and to get out of the house. And I was proud to be able to teach my children about my profession. But I always yearned to get back to my child(ren). This is a very normal response, but so is the opposite.

DOI: 10.1201/9781003406907-12

Some postpartum mothers feel a strong pull to get back to work, and they feel their mental health is much better when they are not home with their baby all day long. They long to use a different part of their brains, to have a more structured day, to interact with adults, to have a reason to leave the house and put on different clothes. This can be particularly true for veterinary professionals who are used to being busy and productive all day long, used to saving lives and problem-solving. In some cases, babies are colicky and don't sleep or don't want to be set down, and it can be a welcome break to step away and feel like a person separate from just being there to cater to baby's every (seemingly never-ending) need.

For some mothers, the idea of leaving their child to go back to work can be devastating. Particularly for first-time mothers, this can come as a shock. What had once seemed like a foregone conclusion—that a mother would return to work at a certain time after the birth of the baby, with a certain schedule, and with a certain childcare arrangement— now does not feel so certain, or at least it no longer feels so easy or desirable. Some people, even those who had been very career-oriented and always had a strong desire to continue in their profession, can find themselves calling into question their desire to return to work at all. And for some, the decision to return to work becomes one of necessity: for finances or insurance benefits.

Some working parents feel a constant struggle between not wanting to leave their children but wanting to go back to work. Particularly in the United States, many people feel that they have not had enough time to heal, bond, or establish any semblance of a routine in their new relationship with their infant before they must return to work. This is a time when sleep, energy, and sanity can feel hard to come by, while the physical, mental, and emotional demands keep growing. On top of this, some individuals are still in the throes of, or just starting to develop, postpartum depression and/or anxiety. In other countries where maternity leave can be extended much longer, a returning parent may feel out of practice and anxious about their knowledge or skill level after their time away from the workplace. Some may feel judged as a parent for returning to the workplace earlier than they had to. All of these reasons make going back to work a very challenging, emotionally charged time in a mother's life, and for their family and friends as well. This chapter will discuss ways to navigate this challenging milestone in the life of a working parent, with emphasis on empathy, grace, flexibility to change plans, the importance of asking for (and offering) help, and banishing guilt. Ways to adjust working responsibilities and schedule will be reviewed, and the myriad childcare options and arrangements will be covered. Pumping at work and feeding arrangements for baby will be discussed.

NAVIGATING THE TRANSITION BACK TO WORK

RETURN-TO-WORK CHALLENGES

New challenges, both emotional and logistical, go along with returning to work—for example, adjusting to a new routine and arranging care for baby. These challenges include the mental and emotional stress of being separated from baby (or child) perhaps for the first time, needing to be mentally alert and awake at work despite significant sleep deprivation, managing the need to pump frequently throughout the day for mothers who feed breast milk, accounting for the increased time involved in preparing baby bottles and other supplies, and for transport to and from childcare for babies who are cared for outside of their home.

Various physical challenges can complicate a working parent's return to work after giving birth as well. The most commonly reported physical health problems are fatigue, urinary problems, back pain, pelvic pain, sore breasts, and sleep deprivation. Those who have had a cesarean delivery may suffer from itching, numbness, or pain at the incision site. The risk of experiencing many of these physical challenges after returning to work is often highest in those who take fewer than six weeks of parental leave. A retrospective study evaluating self-reported physical health in university faculty and staff in the first month back at work after giving birth revealed that those who took more than 17 weeks of parental leave before returning to work had better reported physical health than those who took 6 weeks or less by a factor 14.5.[1] For those parents who return to work while facing their own health challenges and still healing, the physical demands of veterinary practice may delay and/or complicate their recovery.

Along with mental health concerns, physical health concerns in postpartum working parents can contribute to absenteeism, defined in this case as needing to miss work frequently due to physical or mental health issues. They can also contribute to presenteeism, which is defined as "being present at work, but unable to perform one's job." Both absenteeism and presenteeism are detrimental to the working parent and to the workplace. In fact, some research indicates that "pregnancy, childbirth, and postpartum ill health result in the highest costs of absenteeism and of productivity losses for employers."[1] Therefore, it is to the benefit of not only the working parent, but also the employer, to ensure that parents returning to work after giving birth have the support they need. This can include allowing parents to have sufficient time to heal before returning to work and allowing for any accommodations needed to support the physical and mental well-being of parents when they do return.

BOX 12.1   Practice Profile: Going back to work and setting boundaries

Maria Botinas, DVM

I really had it all under control after the birth of my first child. You could always count on me to make every appointment, be overly organized, get to my shifts on time, and I was always present at work. I thought this was easy and I could handle a second baby. I was hit with reality when my second was born.

After my second was born, I felt like I was living in a ball pit. A ball pit where the balls were continuously being thrown for me to catch them, but instead of catching them, I was drowning in them. My colleagues were lucky if I knew what day it was. I was never on time. My life was out of control. On top of having a second child, throw in a cross-country move… you can imagine the stress I was facing.

I knew something had to give. Physically and mentally, I could not continue this way. I was trying to catch the balls but felt I was failing to do so. I had to instill my own boundaries and put myself and family first.

The weird thing about boundaries is they are supposed to make you feel better and less stressed. In actuality, you wind up more stressed and with feelings of guilt when you start to initiate these boundaries. That is, if you let that happen. I had to actively work on this in the beginning.

My boundaries up to this point were basically non-existent. I just went with the flow, was told what to do, where to be, and I just did it. I never looked at what I needed to be fulfilled in life.

Once I realized that I needed to make a change, how did I do it? I made a list of what is important to me. I tried to dive in and discover what I needed to give me serenity and peace in my life. What did I need to get my life back on track?

The day I presented my boundaries to my boss, I wanted to throw up. The knot in my stomach almost made me tear that paper up and throw it away. I never truly advocated for myself during my career. I just thought I wasn't allowed to or it would get shot down. I thought that wasn't how a veterinarian handled it. I went into the meeting knowing that at some point we may need to compromise, but there were some things that I just couldn't budge on.

I spoke about what I needed to continue where I was. What I needed in my life and the reasons behind it. Maybe I overshared. That is something I used to do, but for me, the word vomit just kept coming to the point where they stopped me. Just as I expected, we compromised on some stuff, but what I truly needed, they agreed to.

This was the beginning of advocating for myself, and it is something that I try to teach to and instill in other veterinarians and others in general. Since that day, while I may not have everything under control, I at least have a foundation. I know that I will be able to drop my children off, that I will be able to make their soccer games, and that I will be an integral part of their childhood.

PREPARING TO RETURN TO WORK

Preparing to go back to work will be most successful with careful planning and preparation ahead of time. This can include:

- Stocking up on all needed pumping supplies, bottles, formula, and/or baby food;
- Having a framed photo of baby to bring to work;
- Packing healthy snacks and water for the workday;
- Meal prepping some frozen dinners (or having others do it) that can be heated quickly;
- Buying new work clothes that inspire comfort and confidence.

Planning to start work in the middle of the week (or even a Thursday or Friday) so that the first week is shorter can reduce stress as everyone adjusts to the separation. As part of this, it may be possible to do some shorter days with the baby's childcare provider even before work starts to help reduce the anxiety everyone feels on the first day of work.[2]

It is also important to set boundaries around work schedules, adding extra responsibilities, and the like. In veterinary medicine, this can mean being clear about the time after which an employee cannot stay at work, scheduling any needed breaks to pump and not allowing this time to be filled with other responsibilities, and discussing whether lunch breaks need to be used for things like going to visit and/or feed the baby. This may also mean saying no to extra cases, walk-ins, and surgeries that might make for a late night and missed breaks. It can be helpful for the returning mother to communicate with their employer or manager before their return to iron out all these details and agree to check in after the first week or two to make sure the agreed-upon arrangements are happening, and that they are working.

In preparation for the first day, a trial run can be completed a few days before. This can include waking up at the indicated time, doing all the preparatory work to get out the door, feeding the baby, driving to the childcare facility if needed, and then driving to work. This can help to determine if the anticipated wakeup time will be early enough, if changes need to be made to the routine, or if more needs to be prepared the night before. Parents may wish to plan ahead of time how they will check in on baby during the day. This might be a text, photos, or videos from a nanny or babysitter; a video chat with a partner or family member who is with baby; or surveillance video or updates in an app from a daycare facility or an in-person visit on a lunch break. Although the pace of veterinary medicine may not easily lend itself to taking time for these types of updates and regular communication, feeling connected to baby throughout the work day can help improve a working parent's mental well-being. Most likely the need for frequent updates will diminish as everyone settles into a routine and confidence in baby's safety, well-being, and happiness grows.

BOX 12.2   Practice Profile: Self-doubt and returning to work

Sara Wohlhueter-Page, DVM

I believe I had some postpartum anxiety, undiagnosed, after having both of my children. I would also describe myself as generally more anxious than the average person. I think those things may have played a factor in the feelings I had when returning to work after maternity leave. I took approximately three months of leave with each of my two children. During maternity leave, I had been so consumed with my role as mom, that both times when it came time to separate myself from the kids and return to work, I felt out of place. Not only was my brain scrambled from lack of sleep and the other everyday challenges of caring for a new baby, but I felt like I had forgotten how to be a vet. I can distinctly remember not remembering the dose of Rimadyl and having to ask a colleague. At the time I laughed about it with that colleague, as a coping mechanism I think, while deep down I felt like I didn't belong at the hospital and that I didn't have any business being vet. The full body quivering, sweating, anxiety, and self-doubt that were so strong in the first days back thankfully didn't last many days. Knowing how I had felt after returning from my first maternity leave, I requested starting off with routine appointments when returning after having my second child. While all of those same negative feelings did resurface, my imposter syndrome was notably less severe than the first. Being able to start back to work with the successful completion of wellness appointments for young, healthy patients provided me some of the confidence I was so desperately seeking.

CHILDCARE

One of the biggest challenges is deciding who will be with the baby (or child) if/when a parent decides to return to work. For some families, the choice is easy, as there is a parent who will stay home, a grandparent ready and willing to watch the baby, or schedules that can be arranged so that one parent is always home. For many other families, those options do not present themselves. There are a variety of childcare options available for those families to choose from. Following is a discussion of several of the most common childcare options to consider, along with some of the biggest pros and cons of each.

TRADITIONAL DAYCARE (NURSERY)

Modern daycare (often called nursery in the UK and other countries) is most often preschool and childcare combined. Children will learn social skills and age-appropriate academics while playing and being well cared for. There are multiple benefits to this type of care. These include the opportunity for socialization, a planned curriculum, and preparation for elementary school.

Daycare also boasts consistent open hours, safety and security measures for children, regular meals and snacks, and knowing someone is always there to care for the child, even if one provider is sick or on vacation. Financially, daycare tends to be less expensive than hiring a nanny, and the expenses may be partially reimbursable using a tax-free dependent care flexible savings account, if one is offered through a parent's employer. Some employers will also offer a corporate discount through participating daycare centers. For parents who struggle and stress over their child taking a bottle, napping, or eating, daycares often tend to have a special way of helping babies and young children create new habits (at least when they are at school). One can also argue that daycare helps to build a child's immunity to infectious disease.

This, of course, also goes to the top of the list of cons. Babies and children in daycare are often exposed to multiple infectious diseases, which can mean frequent illness, inability to attend daycare, and lost time at work for parents who must care for their child until they are recovered and no longer contagious. It often feels like as soon as one illness resolves, the next one starts, and it can continue this way for the first few years of a child's life in daycare. Over time, children will develop immunity to these infections through their exposure, but it can feel like the illness will never end for parents who are in the thick of it.

Apart from the risk for contagious illness or disease, other cons of choosing daycare include the reality of less personalized care than some other childcare arrangements. Depending on the childcare center, the age of the children, required ratios in a given state or country, 1 teacher might be responsible for as many as 4 to 25 children at one time. This can mean much less one-on-one time, and less holding of babies. It can mean dirty diapers that are not changed right away and children who climb up on tabletops or other high surfaces before a teacher can stop them.

Although there is always some inherent risk of injury with young children, the risk can be higher in any childcare setting in which multiple children commingle (just like the risk of injury in dogs at a dog park.) Most children go through stages of hitting and biting as they become more interactive but still lack emotional management and communication skills. It is therefore not unusual for toddlers to come home with the imprint of another child's teeth somewhere on their body. The risk of more serious injury is possible but, fortunately, very rare.

Depending on the center and the geographical location, daycare costs can vary tremendously. Many centers require parents to pay for full-time care regardless of whether they need it, and payment must be made regardless of whether the child attends, even if the child is sick or the center is closed. For some families, including those of veterinary personnel who are being paid low wages and parents of multiple young children, the cost of daycare is prohibitively expensive. The demand for daycare can lead to long waiting lists. There have been cases where parents have had to be on waiting lists before the birth of their child, or even before the conception of their child.

BOX 12.3   Practice Profile: Arranging childcare and returning to work

Kristen Lewandowski, DVM

If you had asked me five years ago about becoming a mother, I probably would have told you that I wasn't interested, and I would rather focus on my career. But after finding and marrying my soul mate my views on parenthood changed. Then the first time I saw my sweet little Greyson with his big blue eyes, platinum blonde curls, and chubby pink kissable cheeks, I knew that my priorities in life were forever changed.

I took a 12-week maternity leave from work. It was a very intense, sleep-deprived period, but even so, it was an absolutely blissful time of my life spent getting to know my new baby boy. I cherished being present for every moment of his day. We never spent a moment apart. Even while he slept I would gaze at him for hours. Whenever the thought occurred of returning to work, it would bring me to tears. I didn't hate my job, but the prospect of missing out on even a single gaze, smile, or giggle was heartbreaking. I dreaded the idea that he might crawl for the first time or speak his first word and I wouldn't be there to witness it.

On my first day back to work I felt like I was in a daze. The dreaded day had finally come that would separate me from my son for hours, 12 to be exact. When I walked into the building, it was nice to see all the familiar faces of coworkers, clients, and patients. I was happy to get back to the profession that I had worked so hard to be a part of, but I had the constant feelings of fear, sadness and guilt from being away from my sweet helpless baby. Throughout the day, my mind would drift off thinking about how Greyson might be drinking a bottle, playing, or napping and he was doing it without me.

I was lucky enough to have my mother fly in for a few months to watch Greyson after starting back to work so that he wouldn't have to start daycare until six months of age, but again, as the time approached for him to start daycare I was inconsolable. My husband and I searched high and low for the perfect daycare, but such a place did not exist. I had a high standard of care for my perfect baby boy. Even an ivory tower with a self-cleaning Diaper Genie wouldn't have been good enough for my little guy, but ultimately we just had to choose the best of what was available at the time.

We ended up settling on daycare that had small class sizes and a higher teacher to infant ratio than the other daycares. We hoped more attention would make the transition easier for Greyson since he was not used to being left alone. Prior to his start date, we started doing twice weekly visits where the teachers would talk to Greyson and carry him around the infant room for a few minutes. This approach made it so that each consecutive visit seemed to cause him less stress.

I spoke with multiple other moms of infants in Greyson's class to get their perspectives on the school. Their positive feedback helped to ease my mind a bit, but their experiences were not the same as mine. They all told me that their babies went happily to their teacher during morning drop-offs, but Greyson certainly did not. He would cling to me and reach out for me as his teacher took him from me, his blue eyes spilling with tears each time. It broke my heart each morning that I had to leave him there. I would trudge back to my car feeling the weight of the world on my shoulders. I knew that I had to proceed with my day, see all of my scheduled appointments, and call expectant clients with lab results, but all I wanted was to turn around and bring my baby back home with me for all day snuggle and play time.

Each day as I got into my car and pulled away from the daycare, I would call my ever-patient husband and ask him to remind me of the benefits of daycare. He would tell me again how beneficial daycare was for Greyson's socialization, early education, and immunity. I knew that my son would ultimately be better off for the skills he was learning each day, but this offered me minimal comfort when handing him over to a virtual stranger each morning.

My days at work were more stressful than they were prior to my maternity leave. I was breastfeeding which meant that I needed to pump every three hours throughout the day. My employer was great about giving me a private place to pump and allowing me to alter my appointment schedule as needed, but trying to keep my appointments on time so that I could pump during my designated times was a constant stressor. Due to my pump breaks I couldn't see as many appointments as I did previously, which made me feel like I wasn't pulling my weight as a doctor. I had to be more cautious around aggressive pets to avoid injuries that might indicate medications that could affect my son through my breast milk.

Ultimately, I know that my struggle is the same as many other working mothers. I know that when Greyson is a bit older, he will be proud of his mommy, knowing that she is helping animals and their humans whenever she is away. My career will expand his horizons and give him opportunities for learning and growth. I try to remind myself of how fortunate I am to have a wonderful family that I want to spend all of my time with. Before Greyson, time seemed like an unlimited commodity. Now I realize that time is the most precious gift of all and I treasure every smile, every giggle, and every moment that I am fortunate enough to spend with my beloved family.

IN-HOME DAYCARE

In-home daycare can provide many of the benefits of a larger daycare center, but with a smaller, more family-style environment. There are typically a very small number of children present in this setting, which allows for more individualized care and an "at home" feel. This can be more appealing to some parents who want their child to have more personalized attention while still getting the socialization benefits they would have at daycare. For some children, in-home daycare can be like a second home to them, with the same caregiver(s) every day, who become like family members. The cost of this type of care is often less than that of a commercial daycare center. There is still typically a curriculum that is followed, meals and snacks provided, and oversight by local authorities that ensure that caregivers meet basic requirements for nutrition, infectious disease control, safety, and more.

There are potential downsides to this type of care as well. The risk of infectious disease is still present, although it may be lessened because of the smaller number of children and staff. Just as in traditional daycare centers, there is still a risk of injury. Because in-home daycares often have only one or two caregivers, it often means that when a caregiver is sick or decides to take some time off, there is nowhere else for a child to go for daycare. This can leave working parents with a real dilemma. Also, because the care is being given in a private home, the parents will have little control or knowledge of who else is present in the home and potentially interacting with their child. Instruction in an in-home daycare may be less structured than in a daycare center, which may be appealing to some parents but not others. It is possible that some in-home daycares may not be properly licensed, which may mean that the providers have less training and preparation with regard to the education and safety of children. If a provider does not offer information about their licensure and training, parents should ask for it.

IN-CLINIC DAYCARE

A much less common type of daycare is one that is organized at or near the veterinary practice for the child(ren) of the practice's employees. This might be a room or office space in a veterinary clinic building, a house or apartment on or nearby the property, or another arrangement, often with a nanny or babysitter who has been hired by the practice owner or by the parents who need care for their children. In some cases, clinic staff take turns caring for the children. This can have the obvious advantages of close proximity between parent and child and the opportunity to have more frequent contact with the child throughout the workday. This may allow breastfeeding mothers to breastfeed their child instead of, or in addition to, pumping and bottle feeding. For some parents, this scenario is ideal, and it may be less expensive if the cost is split among employees and/or subsidized by the practice owner. The smaller number of children typically present may lower the risk of infectious disease spread as well.

The drawbacks of in-clinic daycare may include a risk of being left without care if the caregiver becomes sick or has an emergency. There may be much less structure in this type of care, a smaller space for the children to move around in, and the risk of exposure to infectious diseases and/or injury from animals. Caregivers in this scenario may or may not have as much training and preparation as a licensed daycare provider, which could result in lapses in safety for children in their care. Some working parents may feel that they have a harder time concentrating with their child close by. Likewise, some young children may feel upset by seeing their parent when they cannot be with them. Employers also need to consider that clinics could face legal problems if their daycare arrangement does not comply with local laws.

NANNY

Having a nanny is one of the most personalized ways to provide childcare. It often means the child stays at home, has less exposure to infectious disease or injury from other children, and receives much more one-on-one attention and care. An in-home caregiver may also offer other services like light cleaning, care for older school-aged children, starting dinner, transportation to and from after-school activities, and more.

Although there can be many benefits to having a child stay at home, there are some potential drawbacks to consider. Although a commercial daycare center or licensing authority would provide background checks and screening of caregivers, the onus is on the parent(s) to screen and select an individual to enter their home and care for their child(ren). When a nanny is sick, has an emergency, or needs time off for another reason, it may leave the family without childcare. Some nannies may not provide as much structure or academic content as a daycare center (although some do), and there may be less opportunity for socialization (apart from any siblings or other children in the home). In addition, having a private nanny in the home is likely the most expensive daycare option (when priced by the hour). Depending on the number of hours worked, a nanny may be considered an employee who needs to have taxes withheld from their check and a portion paid by the employer (the parents) to comply with the law and to allow the nanny to have verifiable income.

NANNY SHARE

As the name implies, this refers to "sharing" a nanny with different families either at the same time or on alternating days. This can reduce the cost of care while still limiting the number of children who commingle and potentially share contagious diseases. It may also allow a child to stay in their own home for daycare and, if children from other families are also present, allow for some socialization.

The cons of this type of arrangement are similar to that of having a private nanny. Because one person is providing care, the families can be left without care if the nanny is sick or otherwise cannot work. If children from more than one family are cared for at the same time, a parent may need to keep their child out of care if they might be contagious to avoid the spread of sickness to other children. The tax concerns of paying a nanny may still be present, and the added factor of multiple families paying the nanny may constitute traditional "daycare," which might have to be licensed to avoid legal penalties.

AU PAIR

An au pair is someone who travels from another country to live with a family and provide an agreed-upon number of childcare hours. They are usually young adults and may be students. They become another member of the family for the time they stay. Before arriving, they have a background check and some childcare training and must be proficient in English (if they are coming to the United States). Because they live in the home, they can provide individualized at-home care for children. While they do receive a stipend for their work, they are also compensated with room and board. As a result, their stipend will often end up being less than what a nanny would charge. Families get to learn about the culture and country from which the au pair comes and help contribute to the au pair's education and cultural experiences.

While au pairs live in the home, they generally do not provide any services other than childcare. They may be taking classes while they are living with the family or otherwise have their own scheduling requirements to work around. There can be fees associated with the agency that finds and screens the au pair, and families may be required to contribute to the au pair's educational expenses in addition to providing room, board, and a childcare stipend. Families must offer their au pair a private bedroom and include them in most family activities, including trips and holidays. Au pairs will also need to have specified paid vacation and time off on the weekends, and their length of stay may be limited to 12 months. Some au pairs may not be qualified to care for babies younger than three months of age or children with special needs.[3]

FAMILY AND/OR FRIENDS

Some parents may choose a more informal arrangement, in which care is provided by a family member such as a grandparent or a close friend. There can be many advantages to this type of care. Having a known and trusted person caring for children can alleviate the stress of having to search for a provider and wonder if they are safe and reliable. In some cases, it can allow for children to be cared for in their own home, which provides comfort, less concern about transportation, and less exposure to contagious diseases. It can promote bonding between family members. Any family traditions and customs can be more easily maintained. The cost may also be significantly less and, in some cases, free.

Disadvantages to this informal approach can include less or no formal training and standardization of the individual providing the care when compared with a daycare provider or nanny; less structure; and the potential for there to be no backup option when the caregiver becomes sick or needs to take time away. There is also the potential for an increased level of conflict or stress, if, for example, the caregiver does not agree with or respect the parents' child-rearing philosophy.

STAY-AT-HOME PARENT

The definition and potential advantages to this arrangement are fairly straightforward: one parent stays at home with the child(ren) to care for them instead of asking and/or paying another individual to do it. The child has the familiarity of a parent and the comfort of staying in their home, and protection from infectious disease exposure; and the parent has the advantage of no commute to or from work and/or daycare. The parental benefits include the aforementioned, plus a flexible schedule, no need to search for or interview candidates, and the financial incentive of not having to pay for childcare.

There is a financial cost, of course, associated with the lost income of the stay-at-home parent. For some individuals, the cost of daycare practically equals their take-home pay, and staying home makes better financial sense. For others, however, the income lost by a parent staying home creates too much of a financial barrier to make this a feasible option. There may be much less of a structured academic component unless the parent has some background or training in traditional daycare and/or they make a concerted effort to include it. Having a stay-at-home parent can make socialization more difficult (both for the child and the parent). Being a stay-at-home parent can be very isolating, stressful, and overwhelming for some parents, and their decision to work outside the home can be just as much of a mental health decision as a financial one.

WORK FROM HOME

While all parents *work* at home, some parents are actively engaged in paid employment, either full time or part time, from home. This has become much more common since the COVID-19 pandemic. Although clinical veterinary practice does not lend itself easily to remote work, there are a variety of settings in which it is possible. These include telehealth, veterinary and medical writing, some industry and sales jobs, toxicology hotline positions, and more. In some cases, having a work-from-home position might also allow for a parent to care for children at home, especially once they are school aged. This type of arrangement can provide many of the same benefits as those listed for stay-at-home parents.

It is important to remember, however, that paid work-from-home jobs still demand a significant amount of time, concentration, and attention during the parent's working hours, whether they be full time, part time, or flexible. It can be very difficult to care for a baby or young child while also trying to focus on paid employment, even when it is done at home. Although there are scenarios flexible enough that would allow for both, trying to do everything at the same time can result in burnout and significant stress. Some remote jobs also require shift work, availability on evenings and/or weekends, and travel. Thus, there are many scenarios in which a work-from-home parent still must use another child-care arrangement during their working hours.

PART-TIME WORK

Working part time, either in or outside of the home, can help some families better manage childcare. Part-time work has the advantages of still providing both the financial benefits and mental health benefits of working, while potentially reducing both the cost of childcare and the time a parent spends away from their child(ren). Some two-parent households may be able to arrange their schedules so that they trade off so one parent is always with the child(ren) while the other parent works. Other families may use one of the other childcare arrangements described previously, but perhaps for fewer hours per week. Other parents may continue to use full-time childcare while using some of their time to accomplish things around the house or—importantly—for self-care.

The downsides to part-time work will vary based on the particular scenario, but the most obvious one would be the financial loss associated with working fewer hours. Part-time work can also mean the loss of employer-sponsored benefits as well. In veterinary medicine, working part time requires setting and respecting boundaries to avoid staying past the end of a scheduled shift or picking up extra shifts if one plans to care for their child(ren) during those times. In addition, not all daycare centers are willing or able to accommodate part-time schedules, especially for infants.

FEEDING: BOTTLES, BREASTFEEDING, AND PUMPING

No matter the type of childcare parents chose, feeding arrangements will need to be made for children who are still bottle feeding. This can be a source of significant stress for mothers who are returning to work, especially if they have had trouble with breastfeeding and/or with introducing a bottle. First, at the risk of repeating a cliché, it is worth saying that "fed is best." Infant formulas provide the necessary nutrition for a baby to grow healthy and strong.[4] So, if the baby drinks nothing but formula, that is okay! If the baby drinks some combination of breastmilk and formula throughout the day, that is okay! If the baby is exclusively breastfed and rejects the introduction of a bottle, one way or another, the baby will get what they need. If baby is being tube fed for any number of reasons, they are being provided what they need in the best way possible for them.

BOTTLES

Bottles end up being a part of most babies' lives, whether they are filled with breastmilk or formula. For babies who are used to breastfeeding, they may object to being fed from a bottle, even if it is filled with breastmilk. For some babies, it can be enough for someone other than their mother to offer the bottle. For others, it is more about finding their preferred bottle and nipple, which will be different for every baby. Some babies will not accept a bottle from any member of their family. Most of these babies will go on to take a bottle when they are in a childcare environment, or when they can tell that their regular source of food is not present. For the few that still refuse, some can be transitioned to a sippy cup (or other feeding alternative), some can be breastfed during their mother's work break, and some will need to wait until the workday is over and then likely want to breastfeed often throughout their time at home. Pediatricians and lactation consultants can help guide parents through feeding concerns, and they can advise parents of the ideal time to introduce a bottle to babies who are exclusively breastfed.

PUMPING AT WORK

Because of the differences in parental leave lengths and the cultural expectations associated with going back to work after having a baby, the need to express breast milk and to pump at work can vary significantly between countries. While many parents in the United States will return to work within 12 weeks, those in other countries may have the option to continue their leave for up to 1 year or more. This may obviate the need to express breastmilk at work, as many babies will be weaned before that time. However, for those who are breastfeeding when they return to work and who choose to express breastmilk for their baby, it can be helpful to know how to prepare and what to expect.

Mothers who breastfeed will often need to express milk during their workday, both to provide milk for their baby to drink, and to help maintain their own milk supply and prevent breast engorgement and mastitis. Pumping can be stressful for many people. There can be concerns over milk supply, proper fit of flanges, trouble with milk letdown, pain with pumping (especially when breastfeeding is just being established and/or when baby is cluster feeding), exhaustion, cleaning and sterilizing pump parts, and more. Add to this the pressures of the work environment:

- Where will I go to pump?
- Is there a refrigerator to store my expressed milk?
- How can I make time in my schedule?
- Will I be looked down on for taking breaks?
- Will my pay/production suffer?
- Will I be able to keep producing enough milk?
- How much extra time will it take to prepare everything and then wash everything?

Additionally, countries vary significantly in terms of the legal protections afforded to breastfeeding mothers. In a study of breastfeeding break policies of 182 member countries of the United Nations, 25% of the countries had no protected breastfeeding or pumping break guaranteed by law, 71% had guaranteed paid breaks, and 4% had guaranteed unpaid breaks.[5] Below is a sampling of protections provided in various countries as they relate to the breastfeeding and/or pumping employee (Table 12.1).

Table 12.1 Breastfeeding/pumping protections by country

Country	Protection provided
United States	Federal Labor Standards Act and PUMP for Nursing Mothers Act: "a reasonable break time for an employee to express breast milk for such employee's nursing child for 1 year after the child's birth each time such employee has need to express the milk" The space cannot be a bathroom, and it must be shielded from view and from intrusion from coworkers and the public. The break must be paid if other types of breaks are paid by the employer. The employee must be "completely relieved from duty" if they are not being compensated.[6,7]
Canada	Employers should "provide for short breaks during working hours to afford nursing employees reasonable time off, without pay, to breastfeed a child and/or to express milk on the work site."[8]
South Africa	Two paid 30-minute breaks must be provided during each workday for breastfeeding or pumping for the first 6 months of the child's life.[9]
United Kingdom	A breastfeeding employee must be given a place to lie down and rest and given a hygienic and private place to express and store milk if needed. Bathrooms are not considered suitable. Each employer should perform an individual risk assessment to determine the employee's needs.[10]
United Arab Emirates	All "able" mothers are required to breastfeed their children until 2 years of age; those who cannot, will be provided a wetnurse. All government offices must provide a nursery to facilitate breastfeeding. Two 30-minute paid nursing breaks are to be provided for the first 18 months after the birth of the baby.[11,12]
The Philippines	Employers must provide a lactation station at work; a minimum 40-minute lactation break must be provided every 8 hours.[11]
Greece	"Breastfeeding leave": a breastfeeding mother is offered a choice of extension of maternity leave or working fewer hours for the same pay; she must be allowed to not work night shifts.[11]
Argentina	Two daily 30-minute breastfeeding breaks until the child is 1 year old.[11]
China	Employee cannot be terminated due to breastfeeding; one hour break must be provided each day for pumping/nursing until the child is one year old.[13]
India	Maternity Benefits Act: employers must provide breaks for mothers to express breast milk for the first 15 months after birth. Breaks are fully paid; duration is not stated. If more than 50 employees are present, the employer must provide a "creche," or on-site daycare, where employees can go and nurse up to 4 times daily.[14]

BOX 12.4 Practice Profile: Pumping in solo equine practice

Stacey Cordivano, DVM

As told to Emily Singler

Dr. Stacey Cordivano is a mom of two boys, equine veterinarian, practice owner, host of The Whole Veterinarian podcast, and leader in the movement to improve equine veterinarian retention and the overall happiness and wellness of veterinarians. She has owned her own practice for 12 years, which included both of her pregnancies. She describes pumping while managing her caseload and caring for a newborn baby as one of the most stressful parts of being a working mom in veterinary medicine.

Before giving birth to her first child, Stacey reports feeling prepared "in theory" for the idea of breastfeeding and pumping, and she knew that it would be difficult. She had a relationship with a lactation consultant, who provided her with a guide to going back to work that included topics such as how much and how often she would need to pump. She understood that she could make her own schedule and that she didn't have to stick to the rigid timeline of only pumping when the baby would eat.

As a solo practitioner with no relief coverage, Stacey went back to work 11 days postpartum after the birth of her first child, with her mom and her newborn baby coming along for the ride. After a few weeks of doing that, she decided it wasn't working. It was too stressful to try to care for her baby, then treat her patients while listening to the baby cry, and feeling split in so many directions. So she transitioned to scheduling all of her appointments in the middle of the day so she could leave her baby at home for a short period of time to see her patients. This led to always feeling rushed and stressed.

Pumping while being on the road was also very hard. The wearable pumps were not yet out on the market, and she remembers her pump always needed to be plugged into the car. She also had a plug in cooler to keep her milk cold, and it drained her vehicle battery more than once. There was no sink, no refrigerator, and little privacy as she often had to pump in the truck with her technician present. She experienced pumping while driving, spilling milk, being rushed and stressed, and all of the negative effects that can have on breastmilk supply.

With the birth of her second child, she was able to arrange for a colleague to provide coverage for her patients for four weeks (she returned the favor for her colleague's maternity leave). Having gained experience from her first pregnancy, she felt even more prepared this time, so her stress level was lower. She bought a second breast pump to keep in her car, had multiple sets of pumping parts, and used the hands-free cups to help make pumping while driving a little easier. She also set aside more time in her schedule to stop and pump instead of always being in a hurry. Taking these breaks while pumping helped with her milk letdown. She remembers that all these changes had a huge positive effect on her milk supply.

Stacey's advice for other moms is to arm yourself with knowledge, be prepared, and get realistic advice. Then you can decide what works for you. She also feels that looking into the wearable breast pumps may be a great option for moms on the go. While relief veterinarians are becoming more and more prevalent and in-demand on the small animal side, the equine side is lagging behind. And with a reported 40% of AAEP members being solo practitioners per Dr. Cordivano, having more relief equine vets would help more equine vets take the time they need at home to heal and bond with their babies before returning to work, which can reduce their stress levels and better set them up for success as they work to manage motherhood and veterinary practice.

WAYS TO MAKE PUMPING MORE SUCCESSFUL

Pumping at work can feel very daunting, especially when a mother first returns to work and is adjusting to a new schedule and being away from their baby. There are measures they can take before returning to work and, once they have returned to work, to make the pumping experience less stressful and more productive (literally). It can help to start with a good-quality pump. In the United States, most health insurance plans will cover the cost of a breast pump, although they may stipulate which ones they cover and where to purchase them. In other countries, breast pumps are less likely to be universally covered by insurance and may need to be purchased, borrowed, or rented. Features to look for in a pump include portability (how big and heavy is it), efficiency (how well/how quickly does it work), noise level (is it very loud), power source (does it require a cord that is plugged into the wall, does it use a battery, or is it operated manually), and discretion (can it be worn under clothing without being as visible). Fortunately, there are a variety of pumps to choose from, and many sources of feedback from mothers who have used them. Some mothers will choose to purchase two pumps—either because they want to have one that they keep at work or in their vehicle, or because they want to have a wearable pump as well as a more traditional pump.

It can be very helpful to start using the pump well before returning to work. This is important to become familiar with the workings of the pump and how long a typical pump session will last, to become comfortable with the cleaning and maintenance of pumping parts, to have expressed milk given in bottles to baby before returning to work, and to stock a freezer with expressed milk to have ready for the first few weeks back at work. Some experts recommend starting to pump around four weeks after birth once the milk supply is established.[15] In many cases, however, pumping starts much earlier than this, because of latching issues, concerns about milk supply, pain with nursing, desire for someone other than the nursing mother to feed the baby, and other reasons. If pumping has not started before the four-week mark, this is a good time to start.

Sometimes it can be helpful to supplement baby with a bottle of formula fed by someone else while the mother pumps, allowing them to store a large quantity of pumped milk in the freezer. Some people will find that they are overproducers and do not need to supplement, but many will find it hard to build up a reserve of expressed milk without feeding at least some formula. Milk production tends to be highest first thing in the morning. If the timing can be accommodated, this is often a great time to try to express some milk. If this timing doesn't work because of baby's sleep (or non-sleep) schedule and the need to get ready for work and out the door, parents should find whatever time works best.

Most pumps will come with instructions about the care and maintenance of the pump parts. This should include how to clean them, if and how often to sterilize which parts, and how often parts should be checked and replaced. Pediatricians and lactation consultants can provide helpful guidance in terms of the need to sterilize (or not). This can be more important in very young babies, those who were born prematurely, and those who are otherwise more immunocompromised. If sterilization is recommended, options include boiling in water on the stove, using a microwave sterilizer, or using a separate bottle sterilizer. Using the sanitize setting on a dishwasher is not quite the same as the sterilizing methods listed above, but it is the next best thing. Some pump parts are not meant to be sterilized, so cleaning instructions should be followed to avoid accidental damage to the product.

In addition to the pump, there are several very helpful accessories to strongly consider purchasing. These include an extra set (or two) of pump parts to switch out so that there is always a set that is clean and dry, milk storage bags and/or bottles (bags are necessary for freezing milk), a hands-free pumping bra or tank top (a practical choice is one that can be worn all day instead of having to put on/take off another accessory while on a pumping break), a tote or carrying bag for the pump (if it is not staying at work), an insulated bag with ice packs to transport expressed milk, and cleaning supplies such as a bottle brush and a drying rack (at least for home use). Ambulatory pumping often also requires a cooler kept in the vehicle to keep milk cold on the go. Some people may find they need to wear breast pads to absorb milk that leaks in between pumping or nursing sessions. Both disposable and reusable varieties are available. Another key "supply" to have ready during pumping sessions is a picture (or 1000) of baby. Looking at pictures of baby has been shown to help with milk letdown, which can result in better, faster milk production.

Multiple different wearable pumps have come to the market in the last few years, which has revolutionized the pumping experience for some moms. Not having to be tied to a bulky pump, and not having to even be sitting down, can provide tremendous freedom and flexibility in many cases. Wearable pumps are more discreet in the sense that they can be worn completely underneath clothing and may make some people feel comfortable enough to not need privacy to pump. Some veterinary personnel choose to complete records, make phone calls, do surgery, and/or see appointments, all while using wearable pumps.

While these pumps represent a significant leap forward in breast pump technology and convenience, they may not be ideal for every parent. They can have a steep learning curve and do not always provide the same level of suction and milk collection when compared to traditional pumps. Some of them will leak if the wearer leans forward or back, and an improper fit can cause pain. Wearable pumps also tend to be significantly more expensive than traditional pumps and are often not covered by health insurance.

Those who plan to pump at work should plan to discuss their pumping needs with an employer or staff before starting to work again. They will need to figure out how often and when they need to pump and set boundaries around those times. They should become familiar with the breastfeeding and pumping protections offered by their country and state/province and research any policies already instituted at their workplace, and then decide what is needed to meet breastfeeding goals. As a general guideline, many people will pump as many times throughout the workday as their baby would need to eat. For very young babies, that may be as often as every two to three hours. Thirty minutes should be allocated for each pumping session to allow for setting everything up, pumping until milk has stopped expressing, and the time it takes to break everything down, store the milk, and clean up. In some cases, it may be acceptable to store used pump parts still assembled in a clean container in a refrigerator in between pumping sessions during a single workday, washing them completely, and letting them dry at the end of the day. A pediatrician or lactation consultant can advise whether this time-saving step is appropriate.

Veterinary medicine is nothing if not unpredictable. Appointments run over, clients arrive late, patients are sicker than expected, walk-ins and emergencies can appear at any time, traffic can delay commutes, and all of these things can encroach upon the time set aside for pumping breaks. It would be easy to decide on a particular day to just work through the break and hope to catch up later. However, if pumping is important to a mother, it must be a priority. That means setting a boundary around scheduled pumping time and then respecting it. If a mother does not respect their pumping schedule, no one else can be expected to respect it.

Some parents have found workarounds that can make it easier to take breaks for pumping sessions when work and life are always busy. They can include pumping in the car on the commute to or from work (even without a wearable pump this is possible), pumping during a lunch break, and pumping while typing records and/or making phone calls. Some small animal veterinarians will take an extra drop-off appointment in place of an in-person appointment to help keep their patient load up, while then taking an appointment slot to pump. As mentioned previously, some people using wearable pumps choose to actively work with patients while pumping. Trying to express breast milk while being focused on work tasks may interfere with milk letdown for some individuals. Personnel who pump should be allowed to sit and pump during their breaks without having any other responsibilities at that time and without being made to feel that they need to "make up" that time in some other way.

It is very important for a mother to take care of their body to have good milk production. Although this can be much more challenging when having to balance work, childcare, and interrupted sleep, milk supply becomes even more important when going back to work. Staying hydrated, eating healthy, nourishing foods, and trying to get as much sleep as possible are all necessary. Just the act of going back to work and all that it entails (a new schedule, separation from the child, trying to manage working and pumping) can create stress, and this stress can cause a decrease in milk supply. If pumping can be consistent, and a mother is taking care of themself (mind and body) and has support, the dip in milk supply will probably only be temporary. A lactation consultant can be a good resource to help navigate this challenging transition.

BOX 12.5   Practice Profile: Pumping while saving lives

Amanda Modes, DVM

When I came back from maternity leave with my second son, I was working daytime emergency. One of my colleagues had also had a baby the day before I did, and we were both pumping at work. The hospital was very accommodating and created a private space on the third floor that was available for pumping. It didn't have any windows, but the lights were fine. They had a decent chair and were willing to put in a desk if we wanted to do work as we pumped. The hospital had a very relaxed culture/atmosphere, and my colleague and I decided we were comfortable pumping in the ER office and would rather work there while we pumped than take time to go upstairs. As a result, by choice, we would often be pumping while writing records, discussing cases, or rounding. We were able to seamlessly work it into our day and preferred to spend the time working so we could leave closer to when our shift finished, rather than take the time to solely pump. We made it clear we were going to pump, and others could come in if they wanted, or wait until we were done. While I never saw or interacted with clients while pumping, I was often working around patients as I was pumping. I can remember writing orders while pumping, triaging patients while pumping, and unblocking a cat while pumping. However, my shining achievement was when, while pumping, I removed a stick stuck between the maxillary PM4s on a dog. I was the only doctor on and pumping at the end of my shift as I finished up charts to prepare to round my next doctor, when the dog was brought in as a stat because the owners and dog were so distressed. I came out of the office, had a tech get me some hemostats, and while another tech restrained the dog, I reached down and pulled the stick out. In a matter of moments, the dog was instantly relieved. I had the techs finish getting vitals on the dog, while someone went to update the owner. I then finished pumping before I went to fully examine the dog and speak with the owner. But in that moment of being able to help a dog and simultaneously pump for my son, I felt like one pretty badass vet and mom!

THE PITFALLS AND JOYS OF BEING A WORKING PARENT

After the anxiety and stress associated with navigating pregnancy and childbirth, caring for a newborn baby, mental health challenges, and then all of the logistics involved with going back to work, there continue to be challenges with raising children and working (whether at home, from home, or outside of the home). This is true in any profession. In veterinary medicine, this stress can be compounded by the long days, the challenging workload, the constant needs of patients and clients, and the type A mindset that has often been instilled in the work ethic of veterinary professionals from their early training. Parenthood, of course, brings its own set of never-ending needs, long hours, and exhaustion that compete with the needs of the workplace in many instances.

The term "work–life balance" is mentioned a lot as some attainable combination of time at work and time pursuing other priorities, whether it be hobbies, time with family, or anything else outside of paid employment. The term is frequently highlighted in job advertisements, with many veterinary clinics touting their support of a work–life balance for their employees. Exactly what this term means will be different for everyone. It is almost never a balance, but rather a push and pull, where different needs and interests take priority at different times, and this may be shifting constantly.

Lublin eloquently coins a new term to better characterize this shift: "work/life sway." With this mindset, she explains, working mothers "deliberately move back and forth between the professional and personal sides of their ... lives." She continues, "Mothers who sway know when to say 'no.' Nor do they hesitate to request help from their spouses and supervisors."[16] Even with this new concept, working and parenting is still not easy. There are always choices to be made, and times when a working mother feels like no matter what they do, they are letting someone down.

Take the example of the sick child. As discussed previously in the section on childcare, sick children are typically not permitted to attend most group childcare settings for multiple reasons. First, the child is likely too weak and requires too much care, and they will not want to be separated from their parents when they are not feeling well. Second, many of the illnesses that plague young children are contagious. Not only do childcare workers not want other children in their care to get sick, but they don't want to get sick themselves. Therefore, the care of a sick child almost always falls on a parent. When there is not a parent who was planning to be home, someone will have to call off from work. This can be particularly distressing for veterinary professionals, who know that calling off from work can result in an increased workload for other employees, lost revenue for the practice, and potentially the inability to care for many of the patients who had been scheduled to be seen that day. There is no perfect answer, other than to acknowledge that every child gets sick sometimes and needs to be cared for at home. Understanding and compassion on the part of employers and coworkers are key to reducing the stress on the working parent who must stay home, and to allowing that parent to thrive in the workplace going forward.

In many cultures and countries, the norm for heterosexual couples has often been that the mother would be the one who would stay home with the sick child, while the father would be expected to work. Although the gap in parental responsibilities between partners has narrowed in many relationships in the Generation X and Millennial generations when compared with the Baby Boomer generation, it is still far from being closed. There is still work to be done on improving gender equality in the workplace, which includes expecting (and encouraging) fathers and other non-birthing parents to bear their share of the childcare load when their child is sick.

Another commonly expressed challenge in working motherhood is guilt. As Lublin puts it, "working-mother guilt persists in American society today because gender role expectations haven't evolved enough." In other words, some mothers feel guilty going to work because they feel they are neglecting other duties that are mainly (or entirely) theirs, as in rearing children. As long as there is a feeling in society that this is what women are expected to do, or at least have a larger role in, many women may continue to suffer from working-mother guilt. Some women, however, have been able to escape this guilt trap and enjoy both their working and personal lives without the angst. In her book *Power Moms*, Lublin explains that many of the corporate executive working moms she profiled "have sought to reject (guilt) as a debilitating waste of energy."[16]

Lublin outlines ten tips for ditching working-mom guilt, not all of which will apply to every family[16]:

1. **"Find and Keep a Great Child Care Provider"**: this can be challenging, but knowing your child is happy, safe, loved, and having fun while you are working can significantly reduce your feelings of guilt. This does not have to be a nanny, either;
2. **"Give Children a Voice in Your Work Life"**: consider letting kids who are old enough help determine important times and events when they really need your attention;
3. **"Arrange Workday Getaways with Your Kids"**: this is hard to do spontaneously in veterinary medicine, but a planned paid-time-off day or regular day off during the week could allow for extra bonding time with children;
4. **"Enlist Extensive Help from Extended Family"**: this may not always be an option, but when it is, it can be a huge help. Sometimes this help comes not from family, but from friends, a parents' group, or other community organization (or whoever makes up your village);
5. **"Carve Out Time for Yourself"**: take time to refresh and rejuvenate yourself away from your other responsibilities;
6. **"Streamline Your Priorities"**: automate and organize as much as you can to reduce time spent on less important tasks;
7. **"Take Strategic Breaks"**: this can be the length of maternity leave you want, a good vacation, a different work schedule, or even some time off between jobs;
8. **"Practice Sway Every Day"**: decide on your priorities and stick to them. Be present in whatever task or situation you find yourself in;
9. **"Support the Stay-at-Home Dad"**: don't micromanage your partner; let them do their share;
10. **"Accept your Imperfections"**: give yourself grace and don't beat yourself up. Focus on quality instead of quantity.

THE MOTHERHOOD PENALTY

According to the American Association of University Women (AAUW), the motherhood penalty is defined as "The phenomenon by which women's (and birthing parents') pay decreases once they become mothers." This penalty results in mothers making about 58% of what men (including fathers) earn. This disadvantage includes unpaid maternity leave and other time out of the workforce, needing to reduce worked hours to accommodate childcare schedules, and lost opportunities in terms of promotions or pay raises. Fathers, on the other hand, sometimes experience the "fatherhood bonus," where their pay tends to increase once they become fathers.[17]

This disparity is rooted in an outdated cultural expectation that each family would only have one financial provider, and that person would always be the father. Women who did work were not expected to be as available or as productive because of their assumed need to bear a much larger share of the childcare responsibilities than their partner. AAUW reports that hiring managers are "less likely to hire mothers than women who don't have kids, and when they do hire a mother, they offer her a lower salary." Even though many families are now dual-income households, the gender role gap has not been completely eradicated. Since many workplaces were not originally designed to accommodate the working parent with childcare responsibilities, the long hours and inflexible schedules can make it difficult for mothers to feel successful and reap the financial rewards if they are still expected to bear the brunt of the childcare responsibilities. Because fathers and other non-birthing parents are often not expected to be as actively involved in childcare, they are rewarded financially when they prioritize work over family responsibilities.[17]

There is much work to be done. Fortunately, however, in veterinary medicine the preponderance of women and mothers in the profession has shone a light on the needs of working mothers (and working parents in general). Flexible working hours and part-time options, although not always available, are much more common than they were previously, and many practices (both large animal and small animal) are now owned by working mothers who have personal experience with the struggles of the working mother and are more open to accommodating their needs. Apart from private practice, many government and industry jobs can also be adaptable to the needs of a working parent, whether through part-time schedules, flexible work-from-home arrangements, and/or better benefits such as paid parental leave and childcare assistance. The key to these measures being successful is for them to be available to *all* parents so that both working parents in a household (if applicable) can use them. Mothers will not fully benefit from these efforts until their partners are also afforded the same options and encouraged to take advantage of them.

IT TAKES A VILLAGE

The phrase "It takes a village to raise a child" comes from an old African proverb. The village helps to "provide a safe, healthy environment for children." This village is comprised of "multiple people (the 'villagers') including parents, siblings, extended family members, neighbors, teachers, professionals, community members and policy makers," who help care for a child either directly or indirectly through support of the parents. In modern Western societies, unfortunately, this village has in many cases become "dissipated and fragmented," as various factors separate and isolate families from each other.[18] These factors can include geographical separation from extended family, long hours at work, and a propensity for self-reliance and individualism that can make asking or giving help feel less socially acceptable. This separation and isolation can create tremendous stress for parents as they raise their children without as much of a support network as families really need to thrive. This can be a big contributing factor in working-mother (and working-parent) guilt.

When feelings of guilt emerge about not being able to "do it all" as a working parent, it can be helpful to remember the words of Reupter et al.: "Even though parents may be a child's primary caregivers, a family does not exist in a vacuum." Families need to have "social connectedness," which is defined as "those subjective psychological bonds that people experience in relation to others," to help foster resilience in children, reduce rates of child mistreatment, and lower levels of stress for parents, among other benefits.[18] On the subject of families, Osher and Osher state that "a family is defined by its members, and each family defines itself."[19] Every parent, and every family, no matter their makeup, needs support.

In studies of some non-Western societies and their views toward families and child-rearing, the concept of the village is much more prevalent. In a study in the Cameroonian Nso, for example, mothers would frequently habituate their children to spending time with other adults outside of their families, starting in infanthood. When asked why they did this, one mother answered, "Because just one person cannot take care of the child throughout." Another mother answered, "Because it is not possible that I can be taking care of him alone. He would be disturbing me most often. It means I will not be able to do any other thing."[20] On Sikaiana, a small Polynesian atoll, families frequently "foster" children, meaning they move children between multiple caretakers, and even between families. In this society, fosterage is seen as a "way to express Sikaiana values of generosity and build social relations."[21] While these examples are not provided to suggest that such drastic changes in child-rearing are necessary, it is hoped that they provide helpful reassurance that parents were not meant to struggle alone.

THE BENEFITS OF BEING A WORKING PARENT

Despite all the work, planning, sacrifice, and disparities outlined above, the benefits of both being a mother (or a parent in general) and working in veterinary medicine—both for the parent and for the child—are many.

Aside from the more easily recognized benefits of having a child and raising a family (love, affection, companionship, a new outlook on life), animal studies suggest that the state of motherhood can have a "neuroprotective" effect on the brain. This effect can delay the typical decline in neurogenesis and protect the brain from the negative effects of stress. There were also fewer amyloid deposits (a contributor to the development of Alzheimer disease in humans) in the brains of older animals that had offspring.[22]

Parents who are already working or planning to work in veterinary medicine are likely well aware of the perks of the profession: studying and improving the welfare of animals and humans, protecting public health, solving complex problems, serving others, and many more. For parents working in clinical practice, this means daily contact with animals, which is what most veterinary professionals dreamed of when growing up. Apart from the animal interactions, working in veterinary medicine allows for interactions with adults, intellectual conversations, structure and order to the day, a sense of accomplishment, and financial renumeration.

Combining the benefits of parenthood and working in veterinary medicine can result in more than just a combination of the individual benefits of each. Although the term "balance" can be a misnomer, being a working parent can provide variety, stimulation, and a sense of pride. Working can provide a much-needed mental health break for some parents who are in the throes of a challenging phase of child-rearing, whether it be sleepless newborn life, chaotic toddlerhood, emotional adolescence, or any stage in between. Stepping away briefly from the endless needs of a child, who depends on a parent for everything, to focus on work for a defined set of hours can help parents who feel they have lost themselves gain a renewed sense of self. Likewise, having a family to come home to can give a working professional a renewed sense of purpose and belonging.

For children, a study by the Harvard Business School that spanned 24 countries showed that women who had working moms when they were growing up were more likely to have jobs, more likely to have a supervisory role, and earned more money than women whose mothers stayed home with them. Men who grew up with working mothers were more likely to spend time doing household chores and caring for family members than those who had stay-at-home moms.[23] Multiple studies determined that "children whose mothers worked when they were young had no major learning, behavior or social problems" as a result of their mothers working, and they "tended to be high achievers in school and have less depression and anxiety."[24]

Children of working moms can learn from their mother's example—taking pride in their mother's career; being proud of their dedication to studying, teaching, and serving others.

None of this is to say that the decision to work outside the home is always the right one for every family, or that children of stay-at-home mothers cannot also be just as successful. It is merely to show that children of working mothers can and do thrive. Although it can be very hard for parents to manage the never-ending demands of their time, energy, and sanity, the rewards can be immeasurable.

WAYS TO SUPPORT PARENTS RETURNING FROM LEAVE

Being a working parent will always be hard. But there are steps that employers can take to help support working parents' physical and mental health and help them be successful in both the workplace and their personal and family life. These strategies can also have the added benefit of improving workplace performance and employee loyalty. Different employees and different workplaces will benefit more from some of these strategies than others. It can be hugely beneficial to survey working parents to see what would benefit them most. Some examples include:[25,26]

- Offering and promoting employer-sponsored mental health resources;
- Providing predictability in work scheduling;
- Offering choice and control of scheduling, possibly including reduced hours for parents first coming back from leave;
- Having a family-friendly culture that is supported and modeled by all who are in a leadership position;
- Offering childcare in the workplace;
- Offering a subsidy or a corporate discount for childcare expenses;
- Subsidizing coverage for emergency backup care;
- Allowing employees to bring their child to work when they have no other childcare options;
- Covering the cost of a SNOO responsive basinet rental for parents of young babies so that everyone can get more sleep;
- Creating a working parents Employee Resource Group;
- Offering paid time off for all employees, which working parents will need to utilize from time to time to care for sick children;

- Offering equal pay for equal work;
- Supporting fathers and other non-birthing parents in their childcare responsibilities (including missing work to care for a sick child, leaving work to pick a child up from childcare or school, and not staying late or attending evening functions that interfere with family life).

TAKE-HOME POINTS

- There is no right or wrong way to feel about going back to work after parental leave, no matter the length of the leave;
- Give yourself grace. You are not the same person (physically, emotionally, or mentally) you were before the birth of your child. Take time to learn about this new person, what they need, and how to prioritize those needs;
- There are a multitude of different childcare possibilities to consider, each with pros and cons. Every family can make the decision that is best for them;
- Breastfeeding and pumping can be very challenging. "Fed is best." Those who choose to continue breastfeeding should learn about the protections afforded them, try to determine what they need, and have a conversation about it with their employer before returning to work;
- Boundaries around pumping at work are essential;
- Work–life balance is a myth. It is never easy but choosing a priority in whatever situation you are in, and aiming to be present there, can help you to "sway" instead of trying to balance everything;
- There is still much room for improvement in terms of the gender role gap in double-income heterosexual households. Working to combat biases around gender and parental roles can help to reduce working-mom guilt;
- All parents deserve more support in the workplace. This can help families thrive and erase the "motherhood penalty";
- Finding a village, and relying on it, is how we were meant to live and raise our families;
- Your children will be (and likely already are) proud of you. The work that you do is good for you, and it's good for them;
- Employers can support working parents by easing their transition back into the workplace after parental leave, meeting their needs to express breast milk if applicable, working to improve gender equality in terms of employee expectations and pay, and finding creative solutions that work for those who would benefit from them.

REFERENCES

1. Falletta L, Abbruzzese S, Fischbein R, Shura R, Eng A, Alemagno S. Work reentry after childbirth: Predictors of self-rated health in month one among a sample of university faculty and staff. *Saf Health Work*. 2020 Mar;11(1):19–25.
2. Burry M. Tips for going back to work after maternity leave. February 1, 2022. Accessed August 17, 2022. https://www.thebalancecareers.com/tips-for-returning-to-work-after-maternity-leave-2062223

3. Bridge USA. Au pair. Accessed March 20, 2023. https://j1visa.state.gov/programs/au-pair

4. Stanford Medicine Children's Health. Infant nutrition. Accessed March 15, 2023. https://www.stanfordchildrens.org/en/topic/default?id=infant-nutrition-90-P02236

5. Heymann J, Raub A, Earle A. Breastfeeding policy: A globally comparative analysis. *Bull World Health Organ.* 2013;91(6):398–406.

6. US Department of Labor. Break time for nursing mothers. Accessed September 21, 2022. https://www.dol.gov/agencies/whd/nursing-mothers

7. US Congress. H.R. 3110- PUMP for Nursing Mothers Act. October 22, 2021. Accessed March 8, 2023. https://www.congress.gov/bill/117th-congress/house-bill/3110/text

8. Peddlesden J. Breastfeeding protection in Alberta. July 2012. Accessed September 21, 2022. http://www.breastfeedingalberta.ca/images/pdf%20files/Protecting_Breastfeeding_in_Alberta_July_2012.pdf

9. Jacobs G. Fighting for fair breastfeeding & expressing break laws in South Africa & across the globe. October 24, 2019. Accessed August 24, 2022. https://www.llli.org/fighting-for-fair-breastfeeding-expressing-break-laws-in-south-africa-across-the-globe-2/

10. Health and Safety Executive. Protecting pregnant workers and new mothers. Accessed September 21, 2022. https://www.hse.gov.uk/mothers/employer/rest-breastfeeding-at-work.htm

11. Manes Y. What are breastfeeding laws in other countries? A breakdown of nursing around the world. October 3, 2016. Accessed August 24, 2022. https://www.romper.com/p/what-are-breastfeeding-laws-in-other-countries-a-breakdown-of-nursing-around-the-world-19434

12. Sadek I. United Arab Emirates: Maternity and paternity rights in the UAE private sector. June 28, 2012. Accessed September 21, 2022. https://www.mondaq.com/employee-rights-labour-relations/183950/maternity-and-paternity-rights-in-the-uae-private-sector

13. Unicef China. 3 family-friendly workplace policies and practices every working parent in China should know. July 19, 2019. Accessed March 20, 2023. https://www.unicef.cn/en/parenting-site/3-family-friendly-workplace-policies-and-practices-every-working-parent-china-should-know

14. Paycheck.in. Breastfeeding. April 2, 2021. Accessed September 21, 2022. https://paycheck.in/labour-law-india/maternity-and-work/breastfeeding

15. Babylist. How to pump at work: Babylist's ultimate guide. June 25, 2020. Accessed September 21, 2020. https://www.babylist.com/hello-baby/pumping-at-work

16. Lublin JS. *Power Moms: How Executive Mothers Navigate Work and Life.* New York: Harper Collins; 2021.

17. American Association of University Women. The motherhood penalty. Accessed September 25, 2022. https://www.aauw.org/issues/equity/motherhood/

18. Reupter A, Straussner SL, Weinad B, Maybery D. It takes a village to raise a child: Understanding and expanding the concept of the "village". *Front Public Health.* 2022;10:756066.

19. Osher TW, Osher DM. The paradigm shift to true collaboration with families. *J Child Fam Stud*. 2002;11:47–60.

20. Otto H. Culture-specific attachment strategies in the Cameroonian Nso. Cultural solutions to a universal developmental task. Dissertation, Osnabrück (2008). Accessed September 14, 2022. https://osnadocs.ub.uni-osnabrueck.de/bitstream/urn:nbn:de:gbv:700-2009050119/2/E-Diss881_thesis.pdf

21. Donner WW. Sharing and compassion: Fosterage in a Polynesian society. *J Comp Fam Stud*. 1999;30:703–718.

22. Conaboy C. *Mother Brain: How Neuroscience is Rewriting the Story of Parenthood*. New York: Henry Holt and Company; 2022.

23. Nobel C. Kids benefit from having a working mom. May 15, 2015. Accessed September 26, 2022. https://hbswk.hbs.edu/item/kids-benefit-from-having-a-working-mom

24. Miller CC. Mounting evidence of advantages for children of working mothers. May 15, 2015. Accessed September 26, 2022. https://www.nytimes.com/2015/05/17/upshot/mounting-evidence-of-some-advantages-for-children-of-working-mothers.html

25. Brower T. Working parents are in crisis: New data and the 5 best responses. June 5, 2022. Accessed February 14, 2023. https://www.forbes.com/sites/tracybrower/2022/06/05/working-parents-are-in-crisis-new-data-and-the-5-best-responses/?sh=71cb8c987a8c

26. Happiest Baby. Snoo corporate rentals. Accessed March 8, 2023. https://www.happiestbaby.com/pages/corporate-rentals

Appendix

Further information for each chapter can be found at the following resources.

1. **Radiation and imaging**
   IDEXX Veterinary Radiation Safety Center: https://lowerthedose.org
2. **Anesthesia**
   American College of Veterinary Anesthesiologists guidelines: https://acvaa.org/wp-content/uploads/2019/05/Control-of-Waste-Anesthetic-Gas-Recommendations.pdf
3. **Infectious disease**
   CDC Zoonotic Diseases: https://www.cdc.gov/healthypets/diseases/index.html
4. **Drugs and chemicals**
   Mother to Baby consultations on risks of medications and other exposures during pregnancy and breastfeeding: https://mothertobaby.org; 866-626-6847
   NIH Lactmed database for safety of drugs during breastfeeding: https://www.ncbi.nlm.nih.gov/books/NBK501922/
5. **Injuries and ergonomic hazards**
   CDC Veterinary Health and Safety: https://www.cdc.gov/niosh/topics/veterinary/physical.html
6. **Stress and fatigue**
   Merck Veterinary Wellbeing website: https://www.merck-animal-health-usa.com/about-us/veterinary-wellbeing-study
   Heartstorming Wellness veterinary wellbeing: https://www.heartstorming.co
   Reviving Veterinary Medicine veterinary wellbeing: https://marieholowaychuk.com
7. **Trying to conceive, fertility treatments, and pregnancy loss**
   Carrot Fertility benefits for employers: https://www.get-carrot.com
   Cofertility egg freezing and donation: https://www.cofertility.com
   National Council for Adoption: https://adoptioncouncil.org
   National Network of Abortion Funds: https://abortionfunds.org
   Office on Women's Health Infertility information: https://www.womenshealth.gov/a-z-topics/infertility
   Pathways to Parenthood for LGBT People: https://www.lgbtqiahealtheducation.org/wp-content/uploads/Pathways-to-Parenthood-for-LGBT-People.pdf
   Pregnancy Loss Support Program: https://www.pregnancyloss.org
8. **Announcing pregnancy at work**
   Reporting pregnancy discrimination in the US: https://www.eeoc.gov/pregnancy-discrimination
   Reporting pregnancy discrimination in the UK: https://www.acas.org.uk/dispute-resolution

Reporting pregnancy discrimination in Canada: https://www.chrc-ccdp.gc.ca/en/complaints/make-a-complaint

Reporting pregnancy discrimination in Australia: https://humanrights.gov.au/complaints

The Women's Bureau: https://www.dol.gov/agencies/wb

9. **Planning parental leave**

Parentaly parental leave planning and coaching: https://www.parentaly.com

Center for Parental Leave Leadership: https://cplleadership.com

The Expecting Entrepreneur: https://www.amazon.com/Expecting-Entrepreneur-Parental-Planning-Employed-ebook/dp/B09G9L54HK

10. **The fourth trimester and parental leave**

Be Her Village registry for pregnancy and postpartum services: https://behervillage.com/

A Mother's Guide to the Fourth Trimester: https://www.cuimc.columbia.edu/news/mothers-guide-fourth-trimester

CDC Hear Her website: https://www.cdc.gov/hearher/maternal-warning-signs/index.html?s_cid=DRH_HearHer_PPP_Ad3

CDC Vaccination Schedule: https://www.cdc.gov/vaccines/schedules/hcp/imz/child-adolescent.html

American Academy of Pediatrics parenting website: https://healthychildren.org/English/Pages/default.aspx

University of Notre Dame Safe Cosleeping Guidelines: https://cosleeping.nd.edu/safe-co-sleeping-guidelines/

La Leche League International breastfeeding support: https://llli.org/

Guide to Formula Feeding: https://fedisbest.org/resources-for-parents/guide-for-mula-feeding-pediatrician-dr-chad-hayes/guide-formula-feeding-healthychildren-org/

The Veterinarian Doula Postpartum course and resources: https://www.veterinariandoula.com

11. **Mental health**

Postpartum Support International online support groups:
https://www.postpartum.net/get-help/psi-online-support-meetings/

Veterinary Hope Foundation: https://veterinaryhope.org

Vets4vets and Support4support: https://vinfoundation.org/resources/vets4vets/

DVM Moms: https://thedvmoms.com

Not One More Vet: https://www.nomv.org

Not Another Vet Nurse: https://notanothervetnurse.wixsite.com/navn

Suicide and Crisis Lifeline: https://988lifeline.org; dial or text 988 in the US

12. **Returning to work**

Returnity Project supporting parents returning to work: https://www.thereturnityproject.com

The Company of Dads: https://thecompanyofdads.com

Chamber of Mothers Collective Movement for Mothers' Rights: https://www.chamberofmothers.com

Index

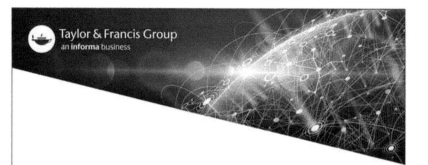